Standards and Best Practice in Absorption Spectrometry

Edited by

C. Burgess
Burgess Consultancy, 'Rose Rae', The Lendings, Startforth, Barnard Castle, Co. Durham DL12 9AB

and

T. Frost
GlaxoWellcome, Dartford DA1 5AH

Ultraviolet Spectrometry Group

Blackwell
Science

© 1999 by
Blackwell Science Ltd
Editorial Offices:
Osney Mead, Oxford OX2 0EL
25 John Street, London WC1N 2BL
23 Ainslie Place, Edinburgh EH3 6AJ
350 Main Street, Malden
 MA 02148 5018, USA
54 University Street, Carlton
 Victoria 3053, Australia
10, rue Casimir Delavigne
 75006 Paris, France

Other Editorial Offices:

Blackwell Wissenschafts-Verlag GmbH
Kurfürstendamm 57
10707 Berlin, Germany

Blackwell Science KK
MG Kodenmacho Building
7–10 Kodenmacho Nihombashi
Chuo-ku, Tokyo 104, Japan

The right of the Author to be identified as the Author of this Work has been asserted in accordance with the Copyright, Designs and Patents Act 1988.

All rights reserved. No part of this publication may be reproduced, stored in a retrieval system, or transmitted, in any form or by any means, electronic, mechanical, photocopying, recording or otherwise, except as permitted by the UK Copyright, Designs and Patents Act 1988, without the prior permission of the publisher.

First published 1999

Set in 10/12pt Times
by DP Photosetting, Aylesbury, Bucks
Printed and bound in Great Britain by
MPG Books Ltd, Bodmin, Cornwall

The Blackwell Science logo is a trade mark of Blackwell Science Ltd, registered at the United Kingdom Trade Marks Registry

DISTRIBUTORS

Marston Book Services Ltd
PO Box 269
Abingdon
Oxon OX14 4YN
(*Orders:* Tel: 01235 465500
 Fax: 01235 465555)

USA
Blackwell Science, Inc.
Commerce Place
350 Main Street
Malden, MA 02148 5018
(*Orders:* Tel: 800 759 6102
 781 388 8250
 Fax: 781 388 8255)

Canada
Login Brothers Book Company
324 Saulteaux Crescent
Winnipeg, Manitoba R3J 3T2
(*Orders:* Tel: 204 837-2987
 Fax: 204 837-3116)

Australia
Blackwell Science Pty Ltd
54 University Street
Carlton, Victoria 3053
(*Orders:* Tel: 03 9347 0300
 Fax: 03 9347 5001)

A catalogue record for this title is available from the British Library

ISBN 0-632-05313-5

Library of Congress
Cataloging-in-Publication Data
is available

For further information on
Blackwell Science, visit our website:
www.blackwell-science.com

Coventry University

This book was published to celebrate the 50th anniversary of the UV Spectrometry Group (UVSG) and is dedicated to all the many people who have given their support to the group over the last 50 years.

Contents

Preface	ix
Part 1: Standards in Absorption Spectrometry	**1**

1 General Considerations on UV-Visible Spectrometry   3
   A.J. Everett, revised by C. Burgess, Burgess Consultancy
   1.1 Introduction   3
   1.2 Radiant energy   3
   1.3 Absorption   5
   1.4 User limitations   7
   1.5 Good spectroscopic practice   15

2 Cell Design and Construction   18
   A.K. Hulme, Optiglass Ltd and A.R.L. Moss, Camspec
   2.1 Introduction   18
   2.2 Cell materials and assembly   28
   2.3 Cell design   34

3 Instrument Design Considerations   43
   T.L. Threlfall, Department of Chemistry, University of York
   3.1 Introduction   43
   3.2 Beam dimensions   44
   3.3 Beam divergence   45
   3.4 Beam masking   46
   3.5 Cell-holders   47
   3.6 Matching cells with instrument   48
   3.7 Self-calibrating instruments   50
   3.8 Diode-array instruments   50
   3.9 HPLC detectors   51

4 Liquid Absorbance Standards   53
   T. Frost, GlaxoWellcome
   4.1 Introduction   53
   4.2 Standards for the 200–400 nm region   53
   4.3 Standards for the 400–800 nm region   61
   4.4 Standards for the entire 200–800 nm region   63
   4.5 Choice of standards   64

	4.6	Commercially available standards	65
	4.7	Conclusions	66
5	Solid Absorbance Standards		69
	D. Irish, Unicam		
	5.1	Introduction	69
	5.2	Glass filters – a historical perspective	69
	5.3	The use of polarizers to determine photometric linearity	73
	5.4	Metal screens	74
	5.5	Sector discs	75
	5.6	Metallic filters	75
	5.7	Light addition methods	77
	5.8	Conclusions and recommendations	78
6	Stray-Light		81
	P. Fleming, Sligo Regional Technical college		
	6.1	Introduction	81
	6.2	Definitions	82
	6.3	Origin of SRE	83
	6.4	Stray radiant energy errors	84
	6.5	SRE reduction	85
	6.6	SRE measurement	89
	6.7	Method comparison	100
	6.8	Conclusion	105
7	Wavelength Calibration		108
	J.G. Vinter, revised by P. Knee, National Physical Laboratory		
	7.1	Introduction	108
	7.2	Line source standards	108
	7.3	Absorption standards	109
	7.4	Other methods	115
	7.5	Conclusion	118
8	Regulatory Overview		120
	J. Hammond, Unicam		
	8.1	Introduction	120
	8.2	Recent history	121
	8.3	Good Laboratory Practice (GLP)	122
	8.4	ISO 9000	123
	8.5	ISO/IEC Guide 25	123
	8.6	International chronology – International Accreditation Conference (ILAC)	124
	8.7	National chronology – United Kingdom Accreditation Service (UKAS)	125
	8.8	ISO 9000 versus ISO/IEC Guide 25	125
	8.9	Qualification and 'ethereal considerations'	127

	8.10	What is the implication for the lab manager of the year 2000?	129
9	Recommended Procedures for Standardization *C. Burgess, Burgess Consultancy*		130
	9.1	Resolution of monochromators	130
	9.2	Wavelength calibration	132
	9.3	Stray-light measurement	135
	9.4	Absorbance standards	140

Part 2: Practical Absorption Spectrometry — 143

10	Absorption spectrometry *A. Knowles*		145
	10.1	Absorption spectrometry in the ultraviolet and visible regions	145
	10.2	The ultraviolet and visible spectrum	147
	10.3	The absorption of radiation	149
	10.4	Molecular structure and absorption spectra	152
	10.5	Quantitative absorption spectrometry	156
	10.6	Measurement of absorption spectra	159
11	Measuring the Spectrum *A. Knowles and M.A. Russell*		165
	11.1	Choice of solvent	165
	11.2	Making a solution	166
	11.3	The cell	167
	11.4	Making the measurement	169
	11.5	Problems and pitfalls	176
	11.6	Cell cleaning	179
	11.7	Accuracy and precision in absorbance measurement	180
	11.8	Difficult samples	186
12	Numerical Methods of Data Analysis *W.F. Maddams*		190
	12.1	Baseline corrections	190
	12.2	Data smoothing	193
	12.3	Multi-component analysis	196
	12.4	Matrix rank analysis	199
	12.5	Spectral stripping and related techniques	201
13	Special Techniques *A.F. Fell, B.P. Chadburn and A. Knowles*		206
	13.1	Derivative spectroscopy	206
	13.2	Difference spectroscopy	211
	13.3	Dual-wavelength spectroscopy	214
	13.4	Densitometry	217

14	Automated Sampling Handling	220
	J.G. Baber	
	14.1 Introduction	220
	14.2 Air-segmented continuous-flow systems	221
	14.3 Flow injection analysis	227
	14.4 Other CF techniques	228
Glossary		230
Appendix		239
Index		243

Preface

The UV Spectrometry Group (UVSG), originally called the Photoelectric Spectrometry Group, was founded in 1948 as a forum for people working with UV–visible spectrophotometry. Some of the background to the formation of the group is given in the Appendix. In the 1980s and early 1990s, the group produced four definitive monographs on UV spectrometry. Volume 1, *Standards in Absorption Spectrometry*, was followed by a similar volume on *Standards in Fluorescence Spectrometry*. Volumes 3 and 4 addressed practical aspects of absorption spectrometry [1–4].

Part 1 of this book comprises a revised and updated version of Volume 1 that takes account of the work that has been carried out and the changes in the regulatory environment since its publication in 1981.

Because Volume 3 is out of print, the opportunity has also been taken to include in Part 2, without revision, some of the chapters on best practice from Volume 3. Although much work, particularly on the chemometric aspects of UV spectrometry, has been published since the first edition of Volume 3, it was felt that it would be useful to draw together the relevant material from that volume.

In the years since the first volumes were published by the UVSG, instruments and instrumentation have progressed far more rapidly than anyone could have predicted. UV spectrometry has become the main detection technique for high-performance liquid chromatography. The introduction of diode array technology and computer-controlled instrumentation has resulted in instruments that are cheaper and more reliable than 20 years ago. The experience of the group in running training courses would suggest that the converse might be true of the instrument operators! The need for standards that provide the 'link with sanity' is as great as ever.

Much of the material from Volume 1 remains intact, and we fully acknowledge the work of the many people involved with that volume. Where only minor changes to the original chapters have been made, we have added the new contributor's name to the contents page. The chapter on stray light has undergone extensive revision and we acknowledge the original work of G.J. Buist, who wrote the original chapter. Much dichromate has flowed since the original chapter on liquid absorbance standards and we would like to acknowledge the work of Elaine Vinter who wrote the chapter on liquid standards in Volume 1. Chapters have been updated where necessary to reflect recent work and the changes in instrumentation. An additional chapter on regulatory compliance has been added and the chapters on cells rationalized into one chapter. The recommended procedures chapter has undergone a significant overhaul and numerous spectra and references have been added. The UVSG references in the Appendix have also been updated and a section documenting the history of the group has been added to these.

We are grateful for the efforts of all the contributors to this book. As with the first volume, all the proceeds from the sale of the book will go to the support of continued research and training in UV spectrometry. The publication of this book is the swan song of the UVSG, as the group has become too small to be viable and is sadly being wound up. The assets are being transferred to the Association of British Spectroscopists Trust to continue the work of the group.

C. Burgess and T. Frost

References

1. Burgess, C. and Knowles, A. (eds) (1981) *Standards in Absorption Spectrometry*. Chapman and Hall.
2. Miller, J.N. (ed.) (1981) *Standards in Fluorescence Spectrometry*. Chapman and Hall.
3. Knowles, A. and Burgess, C. (1984) *Practical Absorption Spectrometry*. Chapman and Hall.
4. Clark, B.J., Frost, T. and Russell, M.A. (1993) *UV Spectroscopy: Techniques, Instrumentation and Data Handling*. Chapman and Hall.

Part 1

Standards in Absorption Spectrometry

1 General Considerations on UV–Visible Spectrometry

1.1 Introduction

Ultraviolet spectrophotometry, as opposed to spectroscopy, has been generally available since about 1943, when it became possible with manual photoelectric spectrophotometers to make reasonably quantitative measurements of the amount of energy absorbed as a function of the wavelength of the incident radiation. Since then, a wide range of manual and recording spectrophotometers has become available, but sadly there is no compelling evidence that the reproducibility of measurements between laboratories approaches that from within a given laboratory. Reasonably competent operators seem able to achieve acceptable precision but often only with rather poor accuracy. Many publications have dealt with this problem which of course is the starting point for this book. The UV Group has played an active part in the quest for the optimum performance of instruments, and a selection of the Group's publications is given in the Appendix and in the references. It is not within our scope to discuss the fundamentals of UV absorption spectroscopy in terms of the electronic phenomenon. The ramifications of quantum mechanics have little impact upon the fingerprint on the front surface of a cuvette. The fundamentals which concern us here are those which bear upon the best use of the available equipment to achieve a spectroscopic measurement. We have assumed that readers will have a spectroscopic background and that the following notes will merely serve to jog the memory as well as bring to mind key references [1–10].

1.2 Radiant energy

Three properties of electromagnetic radiation are necessary to specify it. The quantity or intensity is specified in units of energy or power. The quality is defined by the frequency or vacuum wavelength. Finally, the state of polarization should be specified.

1.2.1 Wavelength

In general, the frequency of UV radiation is too high for direct measurement (about 10^{15} Hz), so that experimental measurements must be in terms of wavelength. Frequency is then derived from:

$$\text{frequency} = \frac{c}{\lambda}$$

where c is the velocity of light in a vacuum and λ is the wavelength of the radiation.

It is important not to confuse frequency with wavenumber. The latter is the number of wave maxima per unit length, being given by:

$$\text{wavenumber}\,(\text{cm}^{-1}) = \frac{10^7}{\lambda(\text{nm})}$$

and although unlike wavelength it is directly proportional to energy, it has no particular spectroscopic significance. Visible light is generally considered to extend from 370 nm to 680 nm, and the near-UV region from 200 nm to 370 nm. Like the other defining wavelengths, the 200 nm limit is arbitrary in that for many instruments, particularly old ones, the stray-light performance rapidly deteriorates with further decreasing wavelength. Oxygen, gaseous and dissolved, and solvent absorption exacerbate the problem even more in the region below 200 nm.

As the use of halographic gratings in spectrophotometers is now widespread, a lower wavelength limit of 185 nm for solution work might be more realistic. However, special precautions need to be observed when working in this region, see later for more details. In most instances, instrument manufacturers provide spectrophotometers whose precision and accuracy of wavelength read-out are adequate, but this must not dissuade the spectroscopist from simple checks of calibration as systematic errors of several nm are not unknown.

1.2.2 Intensity

It cannot be said that the same confidence in wavelength accuracy applies to the measurement of intensity, the second defining property of the radiation. Fortunately for the UV spectroscopist, absolute light intensity measurement rarely arises. It is the attenuation of the light beam which is of more interest to the majority, who are concerned with absorption spectrometry and, here, intensity is loosely equated to absorbance as defined below. If the need does arise, the absolute intensity of the light beam may be expressed in convenient energy units per unit time. The latter aspect is the subject of more detailed consideration in the companion monograph on fluorescence spectrometry [7, 10, 11].

1.3 Absorption

1.3.1 Attenuation of radiation

When a beam of radiation of specific wavelength impinges upon a substance, the energy associated with the beam may be altered by one of four processes; reflection, refraction, absorption and transmission. Most experimental measurements are concerned with elimination of, or corrections for, effects other than absorption.

Usually such measurements are carried out using a spectrophotometer, more correctly called a spectrometer. This consists of a number of basic elements; a source of radiant energy, a wavelength selection device, a sample compartment and a detector. For more details, the reader is directed to the references, in particular [3, 8, 12–14].

The simplest situation with respect to the intensity of absorption is that in which the system obeys the Beer–Lambert law. In this case, if I_0 is the intensity of a parallel beam of radiation incident normally on a layer of thickness b cm and molar concentration c, the intensity of the emergent beam, I, is:

$$I = I_0 \, 10^{-\epsilon c b}$$

where ϵ, the molar absorptivity (litre mole^{-1} cm^{-1}), is independent of c but is a function of wavelength, λ, temperature and solvent. Of course, this implies that each layer, or indeed each molecule, of the absorbing substance absorbs a constant fraction of the incident radiation. The above equation can be expressed in the more familiar logarithmic form:

$$\log_{10}\left(\frac{I_0}{I}\right) = \epsilon c b$$

or

$$A = \epsilon c b$$

where A is the absorbance of the sample in the beam. The ratio of the light intensity transmitted by the sample to the light intensity incident on the sample is the transmittance T:

$$T = \left(\frac{I}{I_0}\right) \quad \text{and} \quad A = -\log_{10} T$$

Transmittance is usually expressed as a percentage, i.e. $\%T = 100\left(\frac{I}{I_0}\right)$, and this convention will be followed in this book. Spectroscopic nomenclature is a never-ending source of confusion, especially in older works. The conventions followed in this book are detailed in Refs [9, 15].

Absorbance is more simply related to concentration and absorptivity than are I, I_0 or T. Strictly, absorbance is only applicable to solutions, the more general term 'optical density' applying to solids and homogeneous liquids as well. However, absorbance will be taken to be synonymous with optical density for our purposes. The attenuation of a beam of radiation in passing through a sample is due in part to absorption within the sample, and in part to reflection and scatter at the external surfaces. The transmission of the material itself, without the external losses, will be termed the 'internal transmission', and is thus defined as that percentage of the radiant flux leaving the entry surface which eventually reaches the exit surface.

1.3.2 Sources of absorbance error

It is convenient to consider two categories of absorbance error. The first originates with the spectrophotometer and the second directly or indirectly with its use. In practice, this dichotomy is not so clearly defined.

Spectrophotometer limitations

At the outset it is desirable to distinguish between the working definition of transmittance or absorbance and the true transmittance or absorbance as outlined by Jones and Sandorfy (chapter IV in Ref. [1]). Using their terminology, for parallel radiation of intensity I_1 falling normally on a cuvette containing a solvent and a solute:

I_r = reflection losses are at cuvette interfaces;
I_s = scattering losses at cuvette surfaces and from the solution;
I_b = absorption losses by the solvent;
I_a = absorption by the solute.

The true transmittance of the solute is:

$$T = \frac{I_i - (I_a + I_b + I_r + I_s)}{I_i - (I_b + I_r + I_s)}$$

On the other hand, the working definition of transmittance, T', generally using a double-team technique, is:

$$T' = \frac{I}{I_0} = \frac{I_i - (I_a + I_b + I_r + I_s)}{I_i - (I'_b + I'_r + I'_s)}$$

It follows that T and T' are only identical when:

$$I_b + I_r + I_s = I'_b + I'_r + I'_s$$

Deviations from this condition are most likely to occur for a sample with low molar absorptivity and high molecular weight.

Reflection losses

If sample and reference cuvettes are made to a sufficiently high specification, the outer face reflections will cancel. Also, so will the reflections at the liquid-to-fused-silica interfaces if, as is usual in UV spectrophotometry, the solute concentration is very low. To put the matter into perspective, the loss from internal reflections in a fused silica cuvette filled with water is only about 0.4% of the incident light energy at 589.3 nm [16]. Even on passing through an absorption band where the solution refractive index and hence the reflectance loss (see Section 1.4.2) is rapidly changing, the effect on the solute absorbance measurement is exceedingly small, being of the order of 0.001%. Measurable effects arising from refractive index imbalance between reference and sample do arise, but they are essentially of an instrumental nature and may be responsible for some of the difficulties associated with the use of potassium nitrate solutions as absorbance standards where the concentrations are as high as 0.15 M.

Solvent absorption

Usually in UV spectrometry the mole fraction of the solute is so low that $I_b = I'_b$, i.e. the numbers of absorbing solvent molecules in each beam are almost identical.

Scattering losses

Small non-conducting particles will, when present as a cloudy sample, exhibit Tyndall scattering whose intensity is proportional to the fourth power of the frequency. This can give rise to very serious problems which lead to apparent deviations from the Beer–Lambert law, particularly at short wavelengths. Good working practices will reduce the gratuitous introduction of scattering errors, see Section 1.5 for more details.

1.4 User limitations

The following factors are ones which should be considered when attempting to obtain the greatest precision and accuracy from a spectrophotometer. In some instances, the instrument design will dictate procedure; in others the user can have a marked influence on the quality of the result.

1.4.1 Gravimetric and volumetric accuracy

This is not the place to deal with these factors *in extenso*. Suffice to note that in most instances weighing errors certainly ought not to exceed 0.1% and volumetric errors should be little more, unless very small volumes are to be handled, in which case the solvent should be weighed and corrections applied for solvent density. Solvation, particularly hydration, is a frequent source of error associated with the measurement of molar absorptivity. It must also be borne in mind that appreciable temperature changes, besides affecting volumetric equipment, will frequently lead to actual changes in the molar absorptivity of the sample. For example, it has been noted that temperature control to within $\pm 1°C$ is necessary for high accuracy work [17].

Particulate matter can be removed by using a proprietary membrane filter on a plastic hypodermic syringe. The syringe should be washed before use, as it may be coated with a lubricant. Glass hypodermic syringes potentially may introduce fine particles of glass and should be carefully cleaned before being used for transferring solutions. Small bubbles of air adhering to the window surfaces are a source of significant error exacerbated by greasy surfaces but alleviated by solvent degassing and careful manipulation of the solutions.

1.4.2 Cuvette handling

An obvious first requirement of any photometric measurement of solutions is that the effect of the container should be measurable or compensated. Ideally, the sample and reference cuvettes should be optically identical. Apart from the identity of the window geometries and the consistent orientation of the cuvettes with respect to the light beams, it is an elementary requirement that they be clean. Recommended cleaning procedures are given in Chapter 2.

The ratio of the reflected light intensity I_r to the incident light intensity, I, intensity on a surface is governed by the Fresnel relationship:

$$\frac{I_r}{I} = \left(\frac{n_1 - n_2}{n_1 + n_2}\right)^2$$

where n_1 and n_2 are the refractive indexes of the two media. Other than at short wavelengths, the transmittance of empty synthetic fused silica will be governed by reflectance losses. For example, the theoretical transmittance of an empty cuvette at the wavelength of the sodium D line (589.3 nm) is 93.3% ($A = 0.0301$), whereas on filling with water the transmittance rises to 99.6% ($A = 0.018$) because the internal reflection losses are decreased as a consequence of the nearer matching of the refractive index of water to that of fused silica. Note that spectroscopic grade water should always be stored correctly (preferably in glass) to avoid the contamination of plasticizers, antioxidants and other absorbing materials. Ideally, cuvettes should only be handled with tongs or hands covered by clean non-shedding cotton gloves. Additionally, the use of absorbent paper tissues is to be avoided and the optical faces of the cuvettes should be cleaned only with lens tissue.

Accuracy is required in setting the cuvette in the beam [2]. For an absorbing medium of refractive index n and of an angle of a radians from the normal to the cuvette with respect to the incident beam, the fractional error in pathlength, δ, is given by:

$$\delta = \left(\frac{0.0123a}{n}\right)^2$$

For many instruments an absorbance of 2 can be read to at least $0.001\,A$ ($\delta = 0.0005$). Even simpler instruments can approach this precision when properly used in the difference mode. An alignment error of just over $3°$ would introduce an error of this magnitude.

To put the matter into perspective, there would have to be a side play of 1 mm in a 1 cm cuvette-holder to introduce this error. Nonetheless, the hazard of using cuvettes with non-standard external dimensions should be recognized. It is also wise to check that the mounting of short cuvettes is such as to give adequate reproducibility, and that micro-cuvettes and long cuvettes do not attenuate either by vignetting the beam at the front window or by the light beam grazing the cuvette walls. The best practice is not to move the cuvette at all, but transfer solutions to and from it (see below).

In most spectrophotometers, the cuvettes are mounted symmetrically with respect to the focus of the beams and pathlength. Errors arising from beam divergency are usually negligible. For the most precise work with dilute solutions, it is best to use only one cuvette and leave it in position for both solvent and solute readings. A disposable polythene (free from plasticizers and antioxidants) pipette is then used for filling, emptying and washing the cuvette. Alternatively, a flow cuvette sipper system may be used. The solvent spectrum readings are subsequently subtracted from those of the solution. Spectrophotometers interfaced to computers now make this approach quite attractive to the user if not to the cuvette manufacturer. The cuvette 'blank' can be stored so that the experiment time remains the same as the conventional method with cuvettes in both beams. Equipment is available to allow the whole process to be automatically controlled. For those common solvents which are transparent over a large wavelength range (octane, water, ethanol) there is little point in insisting on very close matching of the cuvette lengths. Good window parallelism, composition and surface finish are far more important.

Even with perfectly matched cuvettes, some double-beam spectrophotometers are sufficiently sensitive to small refractive index differences between the contents of the two cuvettes to give rise to absorbance errors. This instrumental effect occurs because the pathlength between cuvette and detector is often long, to limit the effects of scattering and fluorescence, and provides an optical lever which moves the light image on the photocathode according to the refractive index in the cuvette. Unless the photocathode response is absolutely uniform over its surface, the beam movement will appear as an absorbance change. These matters are further discussed in Chapters 3 and 6.

1.4.3 Choice of absorbance

Spectrophotometers do not provide constant precision throughout their absorbance range. Therefore, to achieve the best analytical quantitative performance, the combination of solution concentration and cuvette length should be adjusted to the most precise region of the instrument absorbance scale. To a great degree, this region is governed by the noise characteristics of the detector which, for many recording instruments, is a photomultiplier. However, there are an increasing number of instruments employing photodiodes and photodiode arrays. For a photomultiplier detector the noise, N, is related to the incident light intensity in the form:

$$N = K\sqrt{(I_0)}$$

where K is a proportionality constant. By differentiating the Beer–Lambert relationship and recognizing that noise, the error in measuring intensity, obeys the above relationship, the error P, in measuring concentration can be derived in the form:

$$P = K\left[\frac{\sqrt{(I_0)}A}{10^{A/2}}\right]^{-1} \qquad (1.1)$$

This function has a broad minimum at $A = 0.869$ (13.5% transmittance). In practice, it is probably sufficient to use an absorbance between 0.8 and 1.5. However, if stray-light errors are likely, e.g. when using a strongly absorbing solvent, or the spectrophotometer is being used at the extremes of the range of the grating, prism, detector or source, the higher absorbances should be avoided. For a general discussion of the effect of instrumental noise on the precision of spectrophotometric analyses see Ref. [18].

The choice of optimum conditions is not simple. There are many interdependent variables, e.g. cuvette pathlength, concentration, spectral slitwidth, scattered radiation, scale read-out discrimination. A number of publications deal with the matter in some detail, e.g. Refs [1, 2, 6].

1.4.4 Choice of slitwidth

When we set a monochromator to a nominal wavelength, an approximately triangular intensity distribution of wavelengths emerges. That wavelength range which never contains less than half of the peak emergent light energy is termed the effective spectral slitwidth (ESW). The spectral region isolated, i.e. the width of the image of the exit slit along the wavelength scale, is termed the spectral slitwidth (SSW). The natural bandwidth (NBW) of a band measured in the spectrum of a compound is the width at half the height, measured at infinite resolution. The ratio ESW/NBW decides how closely the spectrophotometer measurement approaches the true height. When the ESW is about an eighth of the NBW, the spectrometer will read out 99% of the true height. A number of workers have computed the effect of finite spectral slitwidth on band shape and intensity. Torkington [19] provides data for both Gaussian and Lorentzian band profiles, as well as the following analytical expression relating ESW/NBW to the fraction of the true height for a Lorentzian profile:

$$F = \frac{2}{q} \tan^{-1} q - \frac{1}{q^2} \ln(1 + q^2)$$

where F is the fraction of the true height at the peak maximum for ESW/NBW = $q/2$.

Although UV absorption bands are usually more Gaussian than Lorentzian, this equation holds to within 0.2% up to ESW/NBW = 0.25, and even at ESW/NBW = 0.5 the error is only 1.6%. It is worth noting that the fall in intensity of an absorption peak with SSW is relatively modest compared with the considerable reduction in noise that ensues.

I_0 in Equation (1.1) is proportional to the square of the slitwidth, and therefore:

$$P = K' \left[\text{ESW} \frac{10^A}{10^{A/2}} \right]^{-1}$$

If we are in the habit of maintaining the effective slitwidth at about 0.5 nm and now increase it to be about equal to the natural bandwidth, e.g. 25 nm, the fall in height of the peak will be about 64% compared with a 50-fold gain in sensitivity. This is an extreme example intended to draw attention to the general phenomenon and would only apply if noise is the sole factor which is limiting the sensitivity. It is worth noting that, for an isolated absorption band, the relationship between peak height and concentration will be linear regardless of the value of the SSW, provided that incident radiation all lies within the wavelength boundaries of the absorption band (zero stray-light). The same is true for overlapping bands of the same component, but of course the observed proportionality constant will not be the true molar absorptivity. This does not apply to band overlaps from mixtures of components where the effect of high ratios of ESW/NBW will lead to apparent deviations from the Beer–Lambert law.

When recording spectra with the intention of obtaining the most accurate value of the molar absorptivity at the peak maximum, it takes little extra effort to measure the absorbance as a function of slitwidth. A typical relationship is shown in Fig. 1.1, where the centre of the plateau is obviously a desirable point at which to make measurements. Excessive zeal in closing slits will lead to diffraction errors which will cause a fault from the plateau readings.

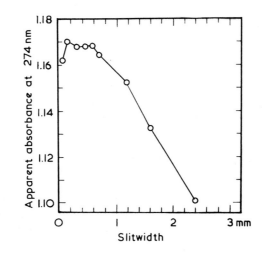

Fig. 1.1 The effect of increasing slitwidth upon the apparent height of a narrow absorption band. The measurements were made on a Beckman Acta CV with the wavelength set at the maximum of the band [16].

1.4.5 *Stray-light*

The wavelength range of a spectrophotometer is largely determined by the energy distribution of the source in relation to the transmission characteristics of the monochromator and the spectral response of the detector. When the spectrophotometer is operated under conditions where any one of these is approaching its wavelength limit, stray-light errors may arise. The fractional effective stray-light of a spectrophotometer is the relative proportion of the detector signal which arises from light scattered within the monochromator (other than that of the nominal wavelength). The stray-light fraction, y, sets a limit to the absorbance range, for clearly the spectrophotometer cannot respond to absorbances higher than $-\log y$. Nearly always stray-light leads to low absorbance values (negative deviations from the Beer–Lambert law) and, in situations where the stray-light fraction is changing rapidly with wavelength, to errors in the wavelengths and shapes of absorption bands.

Beer–Lambert law deviations increase with increasing absorbance. Ultimately, when the sample absorbance is very high, any light transmitted must originate from unwanted radiation and the measured transmittance will approximate to the stray-light fraction:

$$y = \frac{I_s}{I_0 + I_s}$$

where I_s is the intensity of stray-light and I_0 is the intensity of wanted light. I_s sets the dynamic range of the instrument. At short wavelengths (220 nm and below) the incident energy, I_0, decreases continuously with wavelength and y rises. At wavelengths below 200 nm, I_0 will be reduced even further as a consequence of oxygen absorption and, in this region, nitrogen purging of the optical system is desirable.

An important non-instrumental factor which is of great relevance at short wavelengths is the use of absorbing solvents. They lead to a reduction in the proportion of wanted radiation:

$$A_x = \log_{10}\left[\frac{(1-y)T_s}{T'_x y + (1-y)T_s T'_x - y}\right]$$

where A_x = true absorbance of a solute 'x' in a solvent; T'_x = apparent transmittance of solute 'x' measured with solvent reference; T_s = true solvent transmittance, and y = stray-light fraction. When the solvent is transparent ($T = 1$), the equation reduces to:

$$A_x = \log_{10}\left[\frac{(1-y)}{T'_x - y}\right]$$

It is evident that, even for a stray-light fraction of 0.0015 (0.15%), a solvent absorbance of 1 would raise the error in the measurement of an absorbance of 0.8 from 0.4% to 4%. In a poorly maintained optical system, a stray-light fraction of at least 0.01 (1%) may be encountered at wavelengths close to 200 nm, in which case the corresponding absorbance errors become 2.8% and –21.5%, respectively. Ultimately the situation can become so bad that if, optimistically, the monochromator is set at 185 nm there may be so little of this radiation present that the absorbance observations derive from some indeterminate nearby longer wavelength. In these circumstances, one would be better off setting the monochromator to a longer wavelength, even though the molar absorptivity is likely to be lower. To resort to a double monochromator would be an expensive solution to the problem. A cheaper alternative is to avoid working too close to the transmission cut-off of the solvents, as shown in Table 1.1, and to note that stray-light errors are reduced as the absorbance of the solution falls. For example, the errors in the previous example would have been reduced to 1.1% and 10% respectively, if the solution concentration or optical pathlength had been reduced by a factor of 10 to give a nominal absorbance of 0.08. The stray-light problem is discussed in detail in Chapter 6.

Table 1.1 The effective cut-off points for some common solvents. Data from the *UV Atlas of Organic Compounds* [20].

	Effective cut-off wavelength, λ nm	
Solvent	10 mm pathlength	1 mm pathlength
n-Hexane	199	191
n-Heptane	200	189
iso-Octane	202	189
Diethyl ether	205	199
Ethanol	208	196
iso-Propanol	209	
Methanol	211	196
Cyclohexane	212	196
Acetonitrile	213	201
Dioxan	216	208
Dichloromethane	233	
Tetrahydrofuran	238	
Chloroform	247	
Carbon tetrachloride	257	
Dimethyl sulphoxide	270	
Dimethyl formamide	271	
Benzene	280	
Pyridine	306	
Acetone	331	

λ is the value of wavelength at which the transmittance falls to 25% ($A = 0.602$) in the given pathlength, measured using water as the reference

1.4.6 Solvents

A wide range of solvents is available for UV spectroscopy. Since the transmittance of most of these falls at the shortwave end of the range, stray-light errors will become significant if the solvents are used incautiously in this region. As indicated in Section 1.4.5, the effect of a solvent absorbance of 1 in the reference cuvette will be to increase the effect of stray-light approximately ten-fold. A particularly useful set of solvent spectra is provided in the *UV Atlas of Organic Compounds* [20], from which the cut-off wavelengths in Table 1.1 have been derived.

Abandoning the ubiquitous 10 mm cuvette for shorter cuvettes is strongly recommended, though care in cleaning is essential and may be more exacting. The higher solute concentrations rarely give rise to solubility problems but, in regions of high solvent absorption, the usual care concerning the difference in solvent content between the sample and reference cuvettes must be recognized. The single-cuvette technique with subsequent solvent subtraction is probably the best. When working near to the cut-off of ethers and hydrocarbons it should be noted that dissolved oxygen greatly reduces their transmission. Degassing with nitrogen is a simple and undemanding precaution.

1.4.7 Instrument purging

Nitrogen purging the instrument's optical path reduces the interference of oxygen bands below 200 nm and helps reduce the deterioration of the source optics. However, great care must be taken to prevent the distribution of dirt and dust. Many modern instruments have sealed monochromators so that the only accessible part for purging is the sample compartment. Purging is only recommended when critical measurements are being made below 200 nm.

1.4.8 Monitoring procedures

Modern spectrophotometers are long-suffering devices which will continue to give approximate readings long after the uncaring user deserves them. Regular instrument monitoring employing absorbance, stray-light and wavelength checks are relatively undemanding procedures which will expose the false security of blind acceptance. The interval between tests should not be more than 1 week and, for critical work, less. Most spectrophotometer manufacturers now provide simple check procedures. A useful addition is the installation of a clock which records lamp useage hours. Sputtering and evaporation on to the inner surface of the silica envelope of the D_2 arc leads to a steady reduction in the light flux with time. For many arcs, the emission below 220 nm can become seriously low after 100 h of use. Many instruments provide the facility to monitor the hours of lamp usage.

1.5 Good spectroscopic practice

Good spectroscopic practice is a set of pragmatic and practical actions and operations which assist in assuring accurate and reliable measurements. These are discussed in chapter 11 and chapter 9 in Ref. [5] in some detail. However, the following is a list of some of the more important ones.

Ensure that:

(a) The spectrometer is in a proper state of calibration and is well maintained at all times.
(b) The solution concentration is free from weighing, volumetric and temperature errors.
(c) The compound is completely dissolved: ultrasonic treatment as a routine is provident.
(d) The solution is not turbid – filter if necessary – and that there are no bubbles on the cuvette windows.
(e) Adsorption on cuvette walls is not occurring.
(f) The cuvettes are clean and oriented consistently in the light beam.
(g) The reference solvent is subjected to *exactly* the same procedures as the sample solution.
(h) The effective slitwidth is correct for the expected natural bandwidth if absorbance accuracy is important.
(i) Important regions of the spectrum are measured with the sample absorbance lying between 0.8 A and 1.5 A. Adjust the cuvette length rather than the concentration, if practicable.
(j) Stray-light is not responsible for negative deviations from the Beer–Lambert law at high absorbance, particularly if the solvent absorbs significantly.
(k) Regular tests of absorbance and wavelength accuracy are carried out, and check that stray-light is within specifications.
(l) The manufacturer's recommendations are observed.
(m) The environment of the instrument is clean and free from external interference. Particular attention should be paid to electrical interference, thermal variations and sunlight.

References

1. West, W. (ed.) (1956) *Technique of Organic Chemistry, Vol. IX: Chemical Applications of Spectroscopy*. Wiley-Interscience, New York.
2. Lothian, G.F. (1958) *Absorption Spectrophotometry*. Hilger and Watts, London.
3. Bauman, R.P. (1962) *Absorption Spectroscopy*. John Wiley, New York.
4. Marr, I.L. (1974) *Instrumentation for Spectroscopy, in Volume IV of Comprehensive Analytical Chemistry* (ed. G. Svehla). Elsevier Science, Amsterdam.
5. Knowles, A. and Burgess, C. (eds) (1984) *Practical Absorption Spectrometry*. Chapman and Hall, London.
6. Norwicka-Jankowska, T., Gorczyńska, K., Michalika, A. and Wie-

zeska, E. (1986) *Analytical Visible and UltraViolet Spectrometry, Volume XIX of Comprehensive Analytical Chemistry* (ed. G. Svehla). Elsevier Science, Amsterdam.
7. Burgess, C. and Mielenz, K.D. (eds) (1987) *Advances in Standards and Methodology in Spectrophotometry.* Elsevier Science, Amsterdam.
8. Perkampus, H.-H. (1992) *UV–VIS Spectroscopy and its Applications.* Springer, Berlin.
9. Clark, B.J., Frost, T. and Russell, M.A. (eds) (1993) *UV Spectroscopy; Techniques, Instrumentation and Data Handling.* Chapman and Hall, London.
10. Burgess, C. and Jones, D.G. (eds) (1995) *Spectrophotometry, Luminescence and Colour; Science and Compliance.* Elsevier Science, Amsterdam.
11. Miller, J.N. (1981) *Standards in Fluorescence Spectrometry.* Chapman and Hall, London.
12. Burgess, C. (1995) *Radiation Sources in Encyclopedia of Analytical Science*, pp. 3627–3631. Academic Press, New York.
13. Burgess, C. (1995) *Wavelength Selection Devices in Encyclopedia of Analytical Science*, pp. 3631–3638. Academic Press, New York.
14. Burgess, C. (1995) *Detection Devices in Encyclopedia of Analytical Science*, pp. 3638–3643. Academic Press, New York.
15. Spectrometry nomenclature (1973) *Analytical Chemistry*, **46**, 2449
16. Everett, A.J., unpublished work.
17. Gillam, A.E. and Stern, E.S. (1958) *Appendix 3 on Temperature Effects. An Introduction to Electronic Absorption Spectroscopy in Organic Chemistry*, 2nd ed. Edward Arnold, London.
18. Skoog, D.A. and Leary, J.J. (1992) *Principles of Instrumental Analysis*, 4th ed. 131. Harcourt Brace College, Fort Worth.
19. Torkington, P. (1980) *Applied Spectroscopy*, **34**, 189.
20. Perkampus, H.-H. (1992) *UV Atlas of Organic Compounds*, 2nd ed. Verlag Chemie, Weinheim.

2 Cell Design and Construction

2.1 Introduction

A cell is a container for liquid samples especially made for the measurement of absorption or fluorescence in the wavelength range 180–2500 nm, i.e. the ultraviolet, visible and part of the near-infra-red regions of the spectrum. This chapter deals with the types of cell commonly used in currently available commercial instruments, and does not attempt to review all possible types of window material or cell design. Typical cells are shown in Figs 2.1–2.3.

In specifying a cell it is necessary to detail the overall design, material of construction, method of assembly, transmission characteristics of the completed cell, pathlength and dimensional tolerance. Production methods of reputable companies are quality controlled to a much higher standard than was previously available. Surface tolerances are based on other optical definitions, e.g. MIL Specification for scratch and dig. Surface flatness is usually to better than 1 micron across a window face with better than a quarter of a wavelength flatness generally accepted over the working area, since it has a direct bearing on the performance of the cell, specification of the latter is sufficient. Other aspects of the quality of manufacture, e.g. freedom from pinholes, mechanical strength, etc. are even more difficult to specify, and it becomes incumbent on the user to satisfy himself that the cells are of sufficient quality for his purposes.

2.1.1 Definitions

Type of cell

The type of cell refers to its design as far as this relates to the purpose for which it is intended to be used. Various cell types are described in Section 2.1.2.

Grade of cell

The grade of cell refers to the quality of the cell as far as the dimensional tolerances are concerned. This will, in turn, have a bearing on the use of a

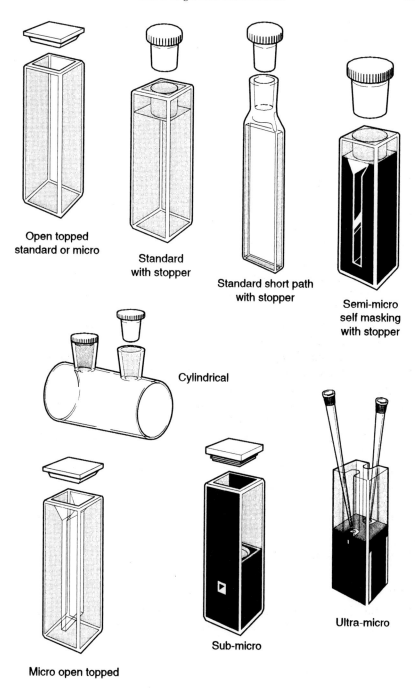

Fig. 2.1 Representative designs of cells.

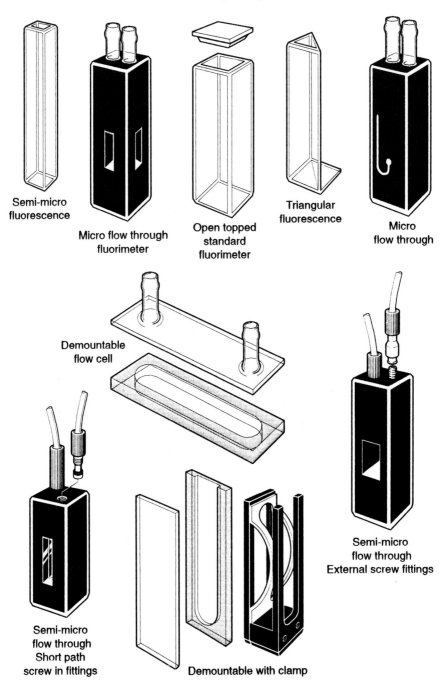

Fig. 2.2 Representative designs of other cells including flow cells.

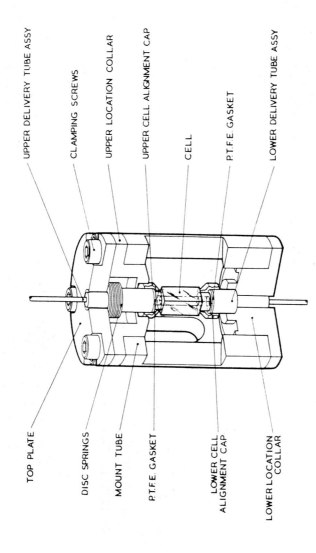

Fig. 2.3 A micro flow cell intended for fluorescence measurements under pressure in an HPLC system. (Diagram by courtesy of Perkin-Elmer.)

Cell windows

A cell must have two optically clear windows or end plates arranged parallel to one another and perpendicular to the measuring beam. A cell for fluorescence measurement generally has additional windows at right angles to the incident beam to transmit the emitted radiation.

Cell walls

The structural parts of the cell that are not required to transmit radiation.

Working area

An area in the windows of a cell which is intended to encompass the measuring beam and which conforms to the tolerances applicable to that grade of cell.

Measuring beam

The instruments referred to in this book are designed to pass an approximately collimated beam of light through the working area of the cell when this is placed in the cell-holder of the instrument. In practice, the beam shape is often an enlarged image of the monochromator exit slit, i.e. an upright rectangle. The beam may well converge, diverge, or converge and diverge by a few degrees during its passage through the cell to the detector.

Fluorescence

Some substances absorb UV or visible radiation and re-emit it at a longer wavelength. Such emission will be termed 'fluorescence' in this book. It is emitted in all directions, but in most commercial instruments is measured only in a direction perpendicular to the incident beam.

Angular deviation

If the two faces of a window are not parallel, or the windows are not parrallel to each other, the measuring beam will be deflected in passing through the cell. This deflection may be in the horizontal or vertical plane, but in most instruments an angular deviation in the horizontal plane will have a greater effect on the accuracy of measurement. The deviation angle varies with the refractive index of the liquid in the cell, and this must be specified when quoting values for the angle.

Dispersion

Any optical defects in the window may cause distortion of the beam as it passes through the cell. Different kinds of refractive faults may give rise to this distortion, which will generally be termed dispersion.

Cell dimensions

These are described in Figs 2.4 and 2.5.

2.1.2 Cell types

The cells dealt with in this book are classified on the lines of the following features:

Static, sampling or flow

'Static cells' are simple containers which are filled and emptied manually, and may or may not be removed from the instrument for refilling. 'Sampling cells' are fitted with tubes so that they can be filled and emptied in situ by pressure or vacuum. Usually they must be emptied as completely as possible before refilling, since they are not designed so that a new sample fully displaces the previous one.

'Flow cells' are intended for continuous flow operation and are designed so that each sample completely displaces the preceding one. This requires that the cross-section of the chamber is a minimum, and that there are minimal dead spaces in the path of the flow. They may be used with a continuously varying sample, as in a chromatography column monitor, or with discrete samples separated by air bubbles, as in an autoanalyser. Examples of these three types are shown in Figs 2.1–2.3.

Absorption or fluorescence

An absorption cell is used to measure the transmission of a liquid, the measuring beam passing straight through it. A fluorescence cell has additional windows, at the bottom or sides, to permit observation at right angles to the incident beam.

Rectangular or cylindrical

A rectangular cell has flat rectangular walls, so the windows and working area are rectangular. The body of a cylindrical cell is formed from a length

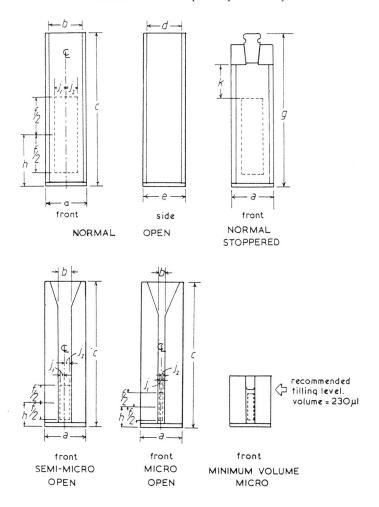

Fig. 2.4 Dimensions and recommended designs for rectangular cells. The broken lines show the working area. (a) External width; (b) internal width; (c) height of cell body; (d) pathlength; (e) external length; (f) height of working area; (g) overall height including stopper; (h) height of centre of working area above base (normally 15 mm); (j) distance of sides of working area from centre-line of cell; (k) distance between top of working area and top of chamber.

of tubing, with a circular window joined to each end, and therefore has a circular working area. The designs of typical rectangular and cylindrical cells are given in Figs 2.4 and 2.5.

Open or stoppered Macro or Standard

An open Macro or Standard cell has a large opening at the top that may be

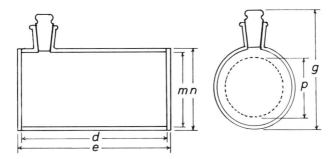

Fig. 2.5 Dimensions and recommended design for cylindrical cells. The broken line shows the working area. (d) Pathlength; (e) external length; (g) overall height including stopper; (m) internal diameter; (n) external diameter; (p) diameter of working area.

covered but not sealed by a lid. A stoppered cell has one or two openings that may be sealed with liquid-tight stoppers. The latter are usually of a standard tapered pattern. Stoppered cells are usually supplied with stoppers, but open cells are not necessarily supplied with lids.

Standard or Macro, Semi-micro and Micro rectangular cells

The best results are obtained with a cell having a large enough working area to accommodate the measuring beam of the instrument under all conditions of its operatioin. The internal width of the cell must be greater than that of the working area, to ensure that the beam cannot be reflected from the walls, or transmitted through them. Such a cell is classed as 'Standard' and a typical example is shown in Fig. 2.4. When less volume of solution is available, a cell can be used in which the volume is reduced by making the chamber narrower. This reduces the width of the working area. The cell is not usually filled to the same depth as the Standard cell, so the height of the working area and its distance above the base are also reduced. Such a cell is termed 'Semi-micro' and an example is shown in Fig. 2.4. They can be used in most spectrophotometers under normal conditions without special precautions, but the operator should check that the beam does not interact with the walls during the measurement.

A further reduction in volume can be achieved by further reducing the working area. This is termed a 'Micro' cell and an example is shown in Fig. 2.4. It will generally be necessary to carefully align the cell in the measuring beam, and to mask the beam to prevent radiation travelling through the walls when transparent (see Section 3.4). Micro cells are available with walls of black glass or black fused silica. These 'self-masking' cells do not require the use of a mask, but must be reproducibly placed in the beam if consecutive measurements are to be compared.

Sub-micro and Ultra-micro rectangular cells

Instrumentation, software and analysis techniques have been improved such that the requirement for cells with volumes in the range from 160 to less than 5 µl are commonplace. The cells which accommodate this requirement fall generally into two groups known as 'Sub-micro' and 'Ultra-micro'.

Sub-micro cells are generally designed for, and refer to, the range from 160 to 10 µl. They minimize sample volume using a small rectangular clear aperture in what would otherwise be a solid black window face. The sample chamber, with an open top, is also rectangular in cross-section and is higher than the aperture so that the meniscus of the liquid is above the aperture and out of the light beam. The fill volume of the sample chamber is slightly more than would be indicated by calculation using the aperture area multiplied by the pathlength. Hence, a 50 µl cell may require 70 µl total volume to compensate for the meniscus effect. Sub-micro cells are designed to minimize the required clearance volume. The solid black surround to the aperture means that special masks are not required, however, orientation and positioning of the cell in the cell-holder becomes more critical for reproducible results. Cells of this design are usually filled with a pipette or syringe.

Ultra-micro cells typically refer to cells with volumes of less than 10 µl. As with Sub-micro cells, they are self-masking and are filled with a pipette or syringe. Because these cells have very small apertures down to 700 microns diameter, signal-to-noise factors also have a bearing on results. The instrument detector technology and beam geometry will determine how critical repeat positioning is in the cell-holder.

To enhance the energy throughput of the sample compartment, there are patented 5 µl Ultra-micro cells available that incorporate a lens within the overall dimension of the cell in front of the sample compartment window. These cells are not suitable for all instruments due to differing optical design and beam parameters, but for those instruments which can accommodate them they do provide a significant increase in energy throughput, therefore reducing the signal-to-noise ratio.

Triangular cells

A fluorescence cell that has three windows, with one at 45° to the others, with the same overall dimensions as a Standard 10 mm cell, is popular for the calibration of spectrofluorimeters. This is illustrated in Fig. 2.2.

Tubular cells

A number of clinical spectrophotometers and fluorimeters will accommodate test-tube-shaped cells. These are mentioned in this book because of their popularity, but are not recommended for accurate measurements.

Demountable cells

Cells for use with samples that contaminate the windows are more readily cleaned if the cell can be partially dismantled. This is particularly useful for short pathlength cells or flow cells with small chambers. One type of static flow cell that is commercially available is shown in Fig. 2.2, together with the spring-clip that holds the window in place. The whole assembly will fit a holder designed for 10 mm cells.

2.1.3 Grades of cells

The Original UV Group Working Party identified three grades of cell based on their dimensional tolerances.

Grade A

These cells will be of the highest quality commercially available for highly accurate work, and specially tested to ensure dimensional accuracy and optical quality. In some cases, e.g. 1 mm pathlength Standard cells, it is impracticable to achieve sufficiently high standards in normal production methods, consequently, no Grade A specification is given for such cells.

Grade B

These are good quality cells for routine use. For precise absorbance measurements they should be calibrated against Grade A cells.

Grade C

These are the cheapest cells and are useful for teaching purposes, or as disposable cells for large-scale routine work and with difficult or dangerous samples.

2.1.4 Special cells

The cell types described in this section are considered to be those in most common use. Even amongst these common types, in some cases it is difficult to guarantee the dimensional tolerances, and the cells are expensive, and have therefore been omitted; e.g. Grade A 1 mm cells were excluded for this reason. Such cells may be obtained from manufacturers by special order, but users are urged to consider alternative ways of making the measurement.

2.1.5 High-accuracy Standard cells

Absorbance measurements of the highest accuracy demand knowledge of the cell pathlength. This can be achieved in one of two ways: (i) by asking the manufacturer to supply a cell that is very close to the nominal pathlength; or (ii) accurately measuring the pathlength of a standard production cell. To illustrate this, the selection of a Standard 10 mm cell required for high-accuracy measurement will be considered. A Grade A cell (see Table 2.1) has a pathlength to better than 10.00 ± 0.01 mm, i.e. within 0.1%. If required, a manufacturer could supply a cell with closer tolerances; this would probably be done by controlling a particular production batch with greater accuracy, and modifying the procedure or selection from a production batch of Grade A cells. As an alternative, the NIST are able to measure the pathlength of production cells at ten points in the height of the cell. These can be purchased, together with the calibration certificate giving the pathlengths to the nearest $0.1\,\mu m$, i.e. 0.01% of pathlength, and the relevant pathlength for the beam height in the user's instrument can be read off from the table. Mavrodineanu and Lazar [1, 2] give descriptions of these cells and the methods used. A similar conclusion is reached for short pathlength Standard cells.

The Grade B specification gives a pathlength of 1.00 ± 0.01 mm, i.e. within 1%. In view of the difficulties associated with the use of cells of such a short pathlength, we do not recommend their use for measurements meriting a greater degree of accuracy than this. However, if more accurate measurements are to be carried out, we suggest that the best approach is to determine the pathlength of a production cell rather than purchase a cell especially made to closer tolerances.

2.1.6 High-accuracy flow cells

The requirements for traceability and accuracy in many production processes, especially in the pharmaceutical industry, has led to a customer demand for flow cells with pathlengths of high accuracy and in matched groups. This applies particularly to tablet dissolution flow cells. Cells are now commercially available with certified pathlengths of less than 1 mm to within 2 microns. These are individually identified with an engraved serial number and their pathlength. The actual pathlength may be determined after assembly by the interference fringe method referred to in Section 9.1.3.

2.2 Cell materials and assembly

2.2.1 Window materials

In choosing window materials, optical transmission in the relevant wavelength range is the prime consideration. Other factors must also be

Table 2.1 Rectangular absorption cells.

Cell type	Pathlength	Working area		Pathlength (d) ± tolerance			External length (e)		Max. internal	Angular deviation of		
		Min. height (f) (j)	Min. distance sides to C.L. cell (j_1 and j_2)	A	B	C	Max.	Min.		A	B	C*
Normal	1	22	3.5	—	0.01	—	3.55	3.10	10.0	—	5'	—
	2	22	3.5	—	0.01	—	4.55	4.10	10.0	3'	5'	—
	5	22	3.5	0.01	0.02	—	7.55	7.10	10.0	3'	5'	—
	10	22	3.5	0.01	0.04	0.30	12.55	12.10	10.0	3'	5'	10'
	20	22	3.5	0.02	0.10	—	22.55	22.10	10.0	3'	5'	—
	40	22	3.5	0.03	0.10	—	42.55	42.10	10.0	3'	5'	—
Semi-micro	5	10	1.75	0.01	0.02	—	7.55	7.10	4.3	4'	6'	—
	10	10	1.75	0.01	0.04	0.30	12.55	12.10	4.3	4'	6'	10'
	20	10	1.75	0.02	0.10	—	22.55	22.10	4.3	4'	6'	—
Micro	5	8	1.0	0.01	—	—	7.55	7.10	2.3	5'	—	—
	10	8	1.0	0.01	—	—	12.55	12.10	2.3	5'	—	—
	20	8	1.0	0.02	—	—	22.55	22.10	2.3	5'	—	—

All dimensions in millimetres

The recommended height of the centre of the working area above the base (h) should be 15 mm in all cases
The maximum height of the cell body (c) is 45 mm, and for stoppered cells, the height including the stopper should not exceed 55 mm
The minimum distance between the top of the working area and the top of the chamber should not be less than 5 mm
The external width is 12.45 ± 0.15 mm for Grade A and B cells, and 12.40 ± 0.30 mm for Grade C cells
* These figures are based on the tolerances of current plastic cells and may require revision as manufacturing techniques improve

taken into account, e.g. the ability to seal the window effectively to the body of the cell; rigidity, to avoid distortion; hardness, to avoid scratching and resistance to solvents and chemical attack. Five types of material are commonly used for cells in the UV and visible regions.

Synthetic fused silica

In order to make silica free of the impurities found in natural quartz, it is prepared by the high-temperature dissociation of purified silicon compounds to give a material transmitting to below 180 nm. It is virtually free of fluorescence.

Fused quartz

Natural quartz can be fused to give a material that is effectively transparent down to 250 nm, but shows absorption below this because of metallic impurities. Fused quartz shows significant fluorescence.

Special optical glass

This is an optical glass with high UV transmission. A cell made of such glass typically has a 75% transmission point at about 310 nm. Borosilicate glasses have slightly better UV transmission characteristics than this, but are not often used for the manufacture of cells. Optical glasses soften and fuse at lower temperatures than fused silica.

Optical glass

Windows of good quality optical glass transmit well throughout the visible region and down to about 350 nm in the near-UV region.

Plastics

At the present time, disposable cells are moulded from polystyrene, methacrylate and other transparent plastics, which give low-cost cells of acceptable optical quality. Many organic solvents attack the material; the surface is readily scratched and can become coated by material from the sample solution. Transmission curves for cells made from these materials are given in Section 2.2.3.

2.2.2 Wall materials

The walls of the cell need not be made from material of the same optical quality as the windows, except in the case of fluorescence cells. However, the thermal expansion coefficient should match that of the window material and, if the chemical resistance of the walls is different from that of the windows, the manufacturer must mark the cell accordingly. In general, wall materials are a lower grade of the same material as the window.

2.2.3 Window transmission

The transmission of a cell can be measured in a spectrophotometer. The apparent fraction of the incident beam that reaches the detector is determined by:

(a) absorption by the window material;
(b) scatter and reflection at all four window surfaces;
(c) dispersion and deviation of the beam due to optical defects;
(d) fluorescence of the window material;
(e) cleanliness of the windows.

Cells made from synthetic fused silica, plastics and some glass materials have low fluorescence, and so (d) is not significant. However, natural fused quartz can have strong fluorescence, and a fused quartz cell when used in the UV region may emit visible light. Item (c) is not important when using good quality cells. With an empty cell, reflection at all four glass–air interfaces will limit the maximum transmission. Two of these reflections are reduced when the cell is filled, but (b) remains significant, for losses at the outer surfaces still limit the maximum transmission to about 92%.

Figures 2.6 and 2.7 show typical transmission curves for cells made from different types of material. The performance of cells in the UV region should be indicated by giving the wavelength at which the transmission of a clean cell filled with freshly distilled water falls to 75%, measured relative to the transmission of air. This wavelength represents the lower limit to the spectral region in which the cell can be used satisfactorily; the cell can be used at shorter wavelengths, but with reduced accuracy.

Extreme caution should, however, be used at wavelengths less than the 75% value. The transmittance of the sample blank decreases rapidly due to absorption by the cell construction material as the wavelength is reduced. Measurements become highly sensitive to instrument wavelength reproducibility and stray-light, until eventually the blank cannot be set to $100\% T$ (0Abs).

2.2.4 Fluorescence of material

Fluorescent energy from windows can cause errors in both absorption and fluorescence spectroscopy. For example, if a sample with high absorbance in the UV is being measured at short wavelengths, fluorescence of the

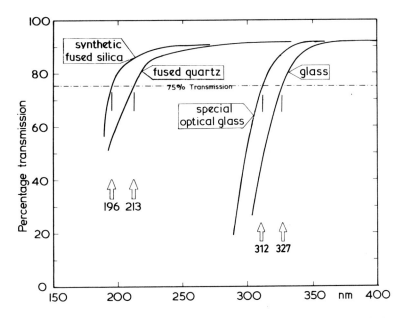

Fig. 2.6 Typical UV transmission curves for cells with windows of synthetic fused silica, fused quartz, special optical glass, and glass. The cells were of 10 mm pathlength and filled with freshly-distilled water. The arrows indicate points of 75% transmission. The poor transmission of the water below 200 nm means that the 75% point of the synthetic fused silica cell is at a longer wavelength than for an equivalent dry cell.

entrance window of the cell, emitted at a wavelength longer than that of the measuring beam, may not be absorbed by the solution and so may constitute a significant fraction of the radiation reaching the detector. This problem is more acute in fluorescence measurements.

The structure of fused quartz has network defects, generally associated with reduced metallic ions that give rise to fluorescence at about 565 nm. Synthetic fused silica is virtually free of metallic impurities and shows negligible fluorescence.

Some glasses show pronounced fluorescence and this, coupled with their UV absorption, makes them unsuitable for most fluorescence measurements.

Typical fluorescence spectra of cells filled with water and measured in right-angle spectrofluorimeters are given in Fig. 2.8.

2.2.5 *Optical specification of the cell*

Several criteria beyond window transmission have to be considered in assessing the optical quality of a cell:

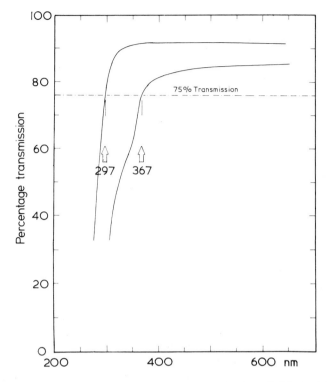

Fig. 2.7 Transmission curves for two makes of plastic cell filled with freshly-distilled water.

(a) closeness to specified pathlength;
(b) parallelism of the inner faces of the windows;
(c) parallelism of the faces of each window;
(d) flatness of the windows over the working area;
(e) freedom from surface blemishes over the working area;
(f) quality of surface polish over the working area;
(g) parallelism of the outer faces of the windows.

Items (a) and (g) can be measured accurately by mechanical means for open types of cell, though short pathlength cells and cylindrical cells present special problems. Items (b) and (c) can also be measured by mechanical methods or, since these defects cause a displacement or deviation of the beam, an optical test can be used.

Item (d) is difficult to check, particularly for the inner faces of windows. However, such defects will cause scattering or dispersion of the beam.

Items (e) and (f) can be assessed by visual inspection, but will also have an effect upon the measured transmission. Thus, a cell that has a poorer

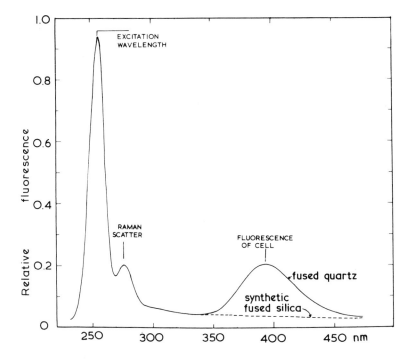

Fig. 2.8 Typical fluorescence emission spectra for cells of synthetic fused silica and fused quartz. The cells were 10 mm square internally, and filled with freshly-distilled water. The spectra were recorded with a Perkin-Elmer Model 204A fluorescence spectrophotometer using 10 mm entrance and exit slits, and an excitation wavelength of 250 nm.

transmission than expected, for the material of its construction, may well have surface defects. It is important to ensure that cells are properly cleaned before any transmission measurement is made, for contamination can have a major effect upon the UV transmission.

2.3 Cell design

2.3.1 *Construction methods*

Fused construction

Most cells that are sold today have walls made from the same physical grade of material as the windows, and direct fusing by the use of heat joins all parts of the cells. Good quality cells should not exhibit distortion even at the edges of the windows. Modern techniques have reduced such distortion to a minimum, and the resulting cell is strong, as resistant to thermal shock as the material of construction, resistant to attack by cleaning agents and solvents, and is easy to clean.

The most satisfactory cells are formed from plates or plates and a bend fused together. The windows are polished and then fused to the polished body. Modern production methods minimize the distortion of the windows and body during the process, but some loss of optical quality around the edges of the windows and slight variability in the pathlength of the cell, depending on the material and cell design, is possible. In addition, fusing quartz and synthetic silica can cause a deposit of amorphous silica to form on the surrounding cold surfaces.

In Standard cells and good quality Semi-micro, Micro, Sub-micro and Ultra-micro cells, distortion of the windows is unlikely to affect the working area, as is specified in Table 2.1. However, Micro cells are sometimes used with beams that fall outside the working area, and then such distortions might have an effect. In Semi-micro and Micro cells made with thick walls, it is imperative that the fusing is carried out such that there are no gaps inside the cell where the windows meet the side walls.

Sintered construction

Some cells are made by using a glass material, which has a melting point lower than that of the walls and windows interposed between the components. The assembly is then heated to 'sinter' the joints together. The lower temperatures used mean that there is no distortion of the windows, but the joints may not be as strong as fused ones, are liable to be pin-holed, and are generally less resistant to chemical attack. Manufacturers should mark cells that are of sintered construction.

Cemented construction

In some cases, components are joined with cement which gives a joint that may be less mechanically strong, less resistant to chemical attack, and more liable to contamination than the window material. As with sintering, this type of production is particularly useful in the manufacture of cells or containers that have large optical windows which may sag or distort under their own weight if taken up to their own higher fusing temperature. Manufacturers should mark cells of cemented construction.

Demountable cells

Cells made with removable windows may be readily cleaned and, if necessary, the windows replaced. Carefully finished surfaces to the cell body, an efficient clamping system and absolute cleanliness are essential to guarantee reproducible assembly of the cell and freedom from leaks. An advantage of this construction is that the body and windows can be of widely differing materials; e.g. special flow cells for chromatograph monitoring are made with a stainless steel body and fused silica windows sealed with PTFE washers. Since the PTFE is relatively rigid, the cell can be reassembled with a reproducible pathlength, though a strong clamping device is necessary to prevent leakage (see Fig. 2.3).

Moulded cells

Plastic cells are generally moulded in one piece. The inner part of the mould must be tapered and so the windows are not parallel. The tolerances of the cell and the optical quality of the window are determined by the quality of the mould, and there are considerable differences in the cells from different manufacturers.

2.3.2 Dimensional tolerances

The preferred pathlengths, dimensions and tolerances for rectangular and cylindrical absorption cells are given in Tables 2.1 and 2.2. The pathlength of the largest rectangular cell is taken as 40 mm, rather than 50 mm as recommended in the DIN specification, for three reasons:

(a) beam divergence in many instruments means that the beam size would exceed the working area over a 50 mm path;
(b) since they will probably be used with samples too dilute to measure in a 20 mm cell, a factor of 2 is more convenient in scaling down the result;
(c) the cell-holders of some instruments will not accept 50 mm cells.

Table 2.2 Cylindrical absorption cells.

Preferred pathlengths	Working area	Pathlength (d) ± tolerance		External length (e)		Max. external diameter (n)	Angular deviation of beam
		Min. diameter (p)		Max.	Min.		
50	16		0.02	52.55	51.0	22.5	5′
100	16		0.02	102.55	102.0	22.5	5′

All dimensions in millimetres
The maximum overall height is 40 mm, including stoppers

Only two pathlengths are proposed for cylindrical cells, as it is only necessary to use these when long pathlengths demand a large working area. A specification for tubular cells is given in Table 2.3.

Three types of fluorescence cell are given in Table 2.4. Other sizes are in use, e.g. 7 × 7 mm, but the adoption of the three sizes listed in Table 2.4 is recommended.

It must again be stressed that the tolerances given in the tables were chosen to be adequate for most purposes at a reasonable cost. Cells can be made to tolerances better than these, and most manufacturers will be able to supply such as special items.

Table 2.3 Tubular absorption cells.

Internal diameter	Internal diameter ± tolerance	External diameter Max.	Min.
10	0.5	12.55	12.10

All dimensions in millimetres

The base may be hemispherical or flattened, but the dimensions of the cell should fall within the above limits from a point 7 mm above the lowest points of the outside surface of the cell

2.3.3 Ease of handling, cleaning and emptying

The dimensional specifications given above allow some variation in the shapes of cells, but other factors should also be taken into consideration:

(a) A slight rounding or chamfer of the external edges is desirable, but since in many cell-holders the cell is located by contact near the junction of windows and walls, excessive rounding will cause poor alignment. Figure 2.9 shows the recommended style of chamfer.
(b) A flat base for rectangular cells will allow them to stand safely on the bench.
(c) The internal corners should be smooth and rounded if possible; there should be no crevices to retain solutions or cleaning agents.
(d) The stoppers should be as large as possible to facilitate filling and emptying. While it is recommended that the socket for the stopper is in the form of a block (see Section 2.3.4), this should be sealed on so as to minimize the amount of liquid trapped in the corners at the top of the cell when it is inverted.
(e) Tall rectangular cells have the advantage of being easy to remove from the cell-holder, but shorter cells are more economical in materials, easier to clean, and have a lower centre of gravity.

Table 2.4 Rectangular fluorescence cells.

Cell type	Pathlength	Working area		Pathlength 1 (b) ± tolerance		Pathlength 2 (d) ± tolerance		External dimensions (a) and (e)	
		Min. height (f)	Min. distance sides to C.L. cell (j_1 and j_2)	B	C	B	C	Max.	Min.
Normal	10	22	3.5	0.05	0.30	0.05	0.30	12.55	12.10
Semi-micro	5	10	1.75	0.02	0.20	0.02	0.20	7.55	7.10
Micro	3	8	1.00	0.02	0.20	0.02	0.20	5.55	5.10

All dimensions in millimetres

The recommended height of the centre of the working area above the base (h) should be 15 mm in all cases

The maximum height of the cell body (c) is 45 mm, and for stoppered cells the height including the stopper (g) should not exceed 55 mm

The minimum distance between the top of the working area and the top of the chamber should not be less than 5 mm

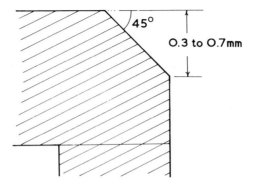

Fig. 2.9 Diagram illustrating the recommended chamfer for the external edges of cells.

2.3.4 Socket and stopper design

Sockets and stoppers of the 'bottle top' design, are recommended for short pathlength cells (i.e. less than 5 mm), as they give an opening that is relatively large compared to the chamber of the cell, for ease of filling and emptying. However, they are rather fragile, and so it is recommended that longer pathlength rectangular cells are fitted with 'block-top' stoppers, as illustrated in Fig. 2.1. The stoppers should have a standard taper of 1 in 10, should be of a standard size, and should be made of the same material as the cell body or of PTFE. In general, PTFE is preferable, as it is less prone to seizing. A sufficient length of stopper should project above the block for its easy removal, and should be shaped so that it can be grasped with the fingertips. It should be borne in mind that PTFE has a much higher coefficient of thermal expansion than glass or fused silica. Since the main function of the stopper is to prevent loss of liquid by spillage or evaporation, it should be airtight.

2.3.5 Fluorescence cell design

The design considerations for absorption cells apply equally to cells used for fluorescence measurements. However, one advantage of the fluorescence technique is that the accuracy of the results is less dependent upon the optical quality of the cells than in absorption photometry. This is because most commercial spectrofluorimeters are arranged so that only the detector views a small central section of the cell. Consequently, tubular cells may be used, though it is important that these have a uniform wall thickness, as this affects the 'lens action' of the cell. The material of construction for a fluorescence cell must transmit both the excitation and emission wavelengths and, since the former is often below 300 nm, glass and plastics are not suitable. Every cell should be checked for intrinsic fluorescence, as variations are found between batches of the same material, particularly in the case of natural quartz. Cemented construction should be avoided since many cements fluoresce.

2.3.6 Sampling cell design

There are many patterns of sampling cell in use, but the pattern shown in Fig. 2.1 is recommended, which has the specification of a Semi-micro cell and will fit in a Standard cell-holder. A 10 mm pathlength cell requires a volume of 1.2 ml to fill it to a level of 5 mm above the working area. The size of the filling tubes is not critical for the sampling mode of operation.

2.3.7 Flow cell design

A typical flow cell is shown in Fig. 2.2. In this case, the dimensions are those of a rectangular Micro cell although, for the greatest efficiency of scavenging and the best time resolution in on-line operation, the volume must be further reduced, with a resulting reduction in working area. The measuring beam must then be suitably masked. In high-pressure liquid chromatography, the highest resolution is required. Cells with a 1 mm diameter working area are often used, and a 10 mm pathlength cell of this dimension has a volume of about 8 µl. Special consideration must be given to the optical design of the measuring instrument used with such a cell.

2.3.8 Flow cell connections

The tubing and connection to the cell should be of the minimum practicable volume; proposals for the connection of tubing are illustrated in Fig. 2.10.

If it were possible to thread the internal bore of a hole in either a glass or quartz cell, then a male connection could be made directly to the cell. As this does not seem to be possible with current technology, two other methods have been developed in an attempt to make a secure connection that will not leak. These are also illustrated in Fig. 2.10.

The first alternative method is achieved by cutting a very accurate thread into the outer wall of a quartz tube. This is then fused directly to the quartz cell with the bore in alignment with the drilled holes in the cell. A female connector is then attached comprising of a flanged tube, which is compressed by a specially machined ferrule, when the outer cap is screwed onto the thread. This has the distinct advantage of accurate hole alignment and virtually zero dead space. If any leak does occur it is immediately visible on the outside of the cell.

The second alternative is to provide a black anodized metal jacket that has two thin walls and a thick block on the top. The top section has M6

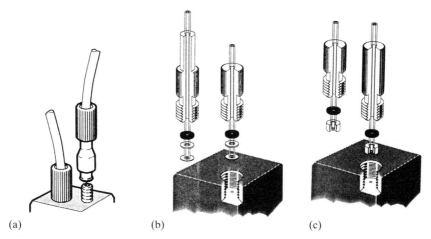

Fig. 2.10 Proposals for the connection of tubing to cells. (Courtesy of Starna.)

holes drilled through it such that when a specially machined glass or quartz cell body is bonded into the jacket the entrance holes in the cell body align with the centre of the M6 holes in the jacket. This allows a male connector with either a flanged fitting or gripper fitting to be screwed into the hole to make a connection when the flange or gripper fitting meets the polished face of the top of the cell body. Unfortunately, any leaks which do occur are hard to spot as they may be well down in the cell-holder. The cell body in some instances may be forced out of the jacket by over-tightening. One major manufacturer who provides a base in the jacket design has overcome this.

Screw connections of both types are particularly useful for use in moving turret or multiple cell-holder instruments used in applications, e.g. tablet dissolution.

2.3.9 Labelling of cells

Three pieces of information should be indelibly marked on the cell in a prominent position outside the working area:

(a) The 75% transmission point in the UV. This should be given as a three digit number, to the nearest 5 nm. The DIN specification recommends that both UV and IR limits should be given to the nearest 10 nm, but we feel that the IR performance up to 1000 nm is not critical for most measurements and is rarely an important factor in choosing a cell. On the other hand, a precise 75% point in the UV is important in the selection of a cell and also gives a good indication of the nature of the window material.

(b) Whether materials differing in nature from the window material have been used in the construction, e.g. glasses of different chemical or physical properties for the walls, or cements or different glasses for the joints. Plastic stoppers or lids are not considered to be part of the cell in this context.

(c) The entrance window should be marked. Putting the above markings on the entrance window above the working area can do this. In the case of plastic cells, this is impracticable and cells are often marked on the base. In this case, an arrow should indicate the entrance window.

In the case of flow and sampling cells, the entry tube of the cell should be marked by an arrow pointing in the direction of flow, placed on the cell body as near as possible to the point at which the tube joins the body. If this is impractical, the tube itself should be marked with an indelible band.

Other information, e.g. the maker's name, can be marked on any face of the cell so long as it does not interfere with the working areas of the windows.

References

1. Mavrodineanu, R. and Lazar, J.W. (1973) *Clin. Chem.* **19**, 1053.
2. *Standard Reference Material 932 – Quartz Cuvette for Spectrophotometry.* NBS Special Publ. 260–32 (obtainable from Superintendent of Documents, US Govt Printing Office, Washington DC 20402).

3 Instrument Design Considerations

3.1 Introduction

This chapter considers features of the design of spectrometers that have a bearing upon the choice and use of cells. Information has been obtained by discussion with seven spectrophotometer manufacturers: Unicam, Beckman, Hewlett-Packard, Hitachi, Perkin-Elmer, Shimadzu and Varian, and by examining current production instruments.

All of the instruments examined accommodate 10 mm Normal cells conforming to the former standard BS 3875, now withdrawn and not replaced. Most modern instruments are under computer control with restrictions on the combinations of operating parameters so that inappropriate selections cannot be made in normal use. As a consequence, it may be more difficult than in previous times to carry out some of the tests and optical examinations discussed below. The main change that has occurred in instrument design over the last 20 years has been the widespread adoption of the diode array spectrometer as an alternative to the traditional monochromator spectrometer. The essential differences, which are shown in Figs 3.1 and 3.2 taken from Ref. [1], are sufficiently large to justify a separate section on diode array instrument characteristics: Sections 3.2–3.7 therefore refer in the first instance to monochromator-based instrumentation with photomultiplier or similar detector.

Spectrometers with Charged Coupled Detectors, CCD, and Fourier transform spectrometers have become the instrumentation of choice in astrophysical applications for low-intensity and high-resolution situations, respectively. In the laboratory, however, these instruments have made little impact, even though two manufacturers have produced dispersive Fourier transform instruments for specialist applications. Such instruments will not be considered further. A section on UV detectors for High-Performance Liquid Chromatography, HPLC, monitoring has been added because of the widespread use of such systems.

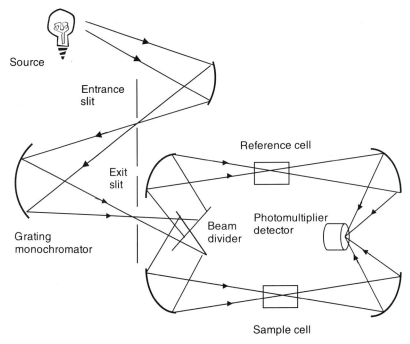

Fig. 3.1 The monochromator precedes the sample area to minimize photolysis of the sample. Rotation of the dispersing element allows the spectrum to be scanned sequentially. Splitting the beam into two identical halves allows ratioing of the transmitted intensity of the beam, with concomitant compensation for solvent and cell in the reference beam.

3.2 Beam dimensions

The light transmitted through a spectrophotometer cell in a monochromator instrument is obtained from a grating, or less commonly, a prism or echelle monochromator. The focus of the light is therefore usually the image of the exit slit of the monochromator. In order that the light can pass through the cell with maximum clearance, the cell is best placed near this focus. The shape of the light beam going through a cell is in consequence approximately an upright slightly bowed narrow rectangle.

Half of the instruments produced before 1980 had beams centred 15 mm above the bottom of the cell, while half had beams centred at 10 mm. Despite the plea for standardization, preferably with the beam centred 15 mm above the bottom of the cell, the range has now widened to 8–15 mm. Cells designed for a beam 8 or 10 mm above the bottom may easily be used in an instrument with a 15 mm beam position by the insertion of a spacer beneath the cell. A 15 mm beam position also allows

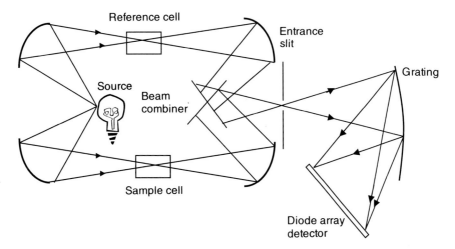

Fig. 3.2 The beam is resolved spacially into a spectrum, which falls upon the elements of the linear photodiode array. It is necessary for the beam to pass through the sample before the monochromator, but this disadvantage is offset by the multichannel advantage of the detector. The monochromator has no moving parts.

the accommodation of such things as magnetic stirrers in the cell below the beam, together with a miniature stirrer driver underneath the cell. The accessibility of the beam within the compartment is rarely of consequence for routine transmission measurements, but the size and design of the cell compartment and the consequent possibility of insertion of additional optical and mechanical components and accessories may be critical for reflection and non-routine measurements, and occasionally for calibration.

3.3 Beam divergence

Among the instruments examined, the common divergence of any ray in the light beam from the axial ray was 3° and the maximum was 5°. For a beam of uniform radiance at all cross-sections, this latter figure would produce an error in the absorbance reading of ±0.2%. This is of similar magnitude to the errors due to cell pathlength (±0.5% in a Grade B cell) and ±0.1% absorbance accuracy, typical of many spectrophotometers. It is recommended that this maximum figure of 5°, equivalent to an f number of about 6, should not be exceeded, which also conforms to the requirements of the International Committee on Illumination [2]. A beam of 5° divergence with an image 0.25 mm wide at the focus will pass through the 2-mm-wide working area of a 20 mm Micro cell without obstruction; it will, therefore, pass readily through most other types of cell.

The beam divergence frequently prohibits the use of rectangular cells with pathlengths longer than 40 mm since the beam cross-section becomes more square away from the focal plane of the slit image. It is recommended therefore that rectangular cells do not exceed 40 mm, and cylindrical cells – better suited to a square beam cross-section – be used for pathlengths of 50 and 100 mm. Beam divergence similarly prohibits the use of long Micro cells, so we have recommended a maximum of 20 mm pathlength for them. Micro cells of 40 mm pathlength may nevertheless be satisfactory in many instruments, but the user should check this by visual inspection, as discussed later. Instrument manufacturers rarely quote beam geometry characteristics in their literature, but are usually willing to provide customers with this information. Otherwise, the geometry can be ascertained by examination of mirror sizes and positions.

3.4 Beam masking

When using Micro cells in some instruments, it will be found that part of the beam falls outside the working area of the cell, or that divergence or misalignment of the beam after it enters the cell causes it to be reflected off the inner faces of the walls. To overcome these effects, the user must arrange a mask to limit the size of the beam.

It is essential, particularly when working with highly absorbing samples, that the measuring beam cannot by-pass the sample by travelling through the walls of the cell; e.g. if 1% of the measuring beam by-passes a sample of absorbance $2A$, the apparent absorbance will be 15% below the true value. The choice of a suitable mask is a compromise arrived at by consideration of the cell dimensions and optical properties of the instrument. The aperture in the mask should be the same size as, or smaller than, the working area of the cell. It is normally placed in front of the entrance window. Ideally, it should be rigidly attached to the instrument, so that movements of the cell-holder or cell will not affect the instrumental baseline. However, in practice, it is often attached to the cell-holder, and care should then be taken that the cell-holder can be reproducibly replaced in the instrument. Another alternative is the use of 'self-masking' Semi-micro cells in which the walls are made of black glass or fused quartz. In this case, care must be taken that the cell is precisely located in the cell-holder and that it lies square-on to the axis of the beam so that the latter cannot be reflected off the inner faces of the walls. It is also essential that the cell is located with the focus of the beam nearer to the exit than the entrance window so that the beam is not diverging as it passes through the cell.

The alignment of the mask, or the self-masking cell, should be checked by visual inspection with the wavelength set at 550 nm and the slits opened wide. Most of the beam should pass through the aperture and, if not, adjustments of the mask, cell or cell-holder must be made. The effect of closing the slit should be noted – the reduction in the size of tie beam, if any, will depend upon the design of the instrument.

The performance of the instrument should then be checked over the operating wavelength range – if the aperture is too small, the instrument may become excessively noisy, or it may become impossible to reach the absorbance zero. In double-beam instruments it may be necessary to attenuate the reference beam in order to get a satisfactory beam balance.

In general, such problems will only be encountered when using Micro cells; the dimensions of Semi-micro cells given in Chapter 2 will allow their use under normal operating conditions without special masking.

3.5 Cell-holders

All the cell-holders examined could accommodate a Normal 10 mm cell. Some instruments have special holders for cylindrical cells and special cells. There is usually a large tolerance on the length, but the cell width may be limited by the width of the holder. A maximum cell width of 12.6 mm has, therefore, been recommended for all rectangular cells.

For accurate re-location of a rectangular cell in a cell-holder, which is particularly necessary for Semi-micro and Micro cells, a corner-locating mechanism is required. While present instruments vary in the means of locating the cell in the corner (a single spring or two perpendicular springs), it is recommended that manufacturers should be encouraged to take the same side of the cell as the reference plane, i.e. the side to the right of the beam looking in the direction of the beam, towards the detector. This means that the external width of the cell is less important, and the thickness of only one wall is critical, hence a smaller internal width for a given working area may be achieved. Due to the problem of friction, care should always be taken when inserting a cell to ensure that it is properly located. Good contact with the cell-holder also ensures better temperature control of the cell. The cell-holder should be suitably recessed so that the working area cannot be scratched as the cell is inserted or removed.

Errors of measurement can arise through mechanical defects of the cell holders:

(a) If the cell is not located squarely in the cell-holder or the cell-holder is not square to the measuring beam, the beam will be displaced on passing through the cell, and the apparent pathlength of the cell will be increased. BS 3875 recommended that the orientation of the face of the cell should be controlled to within 3' in any direction, though in practice this would be difficult to measure. An alignment error of 30' in a 40 mm cell will displace the beam by 0.1 mm, which in most photomultiplier instruments will not cause any serious error, but see Section 3.6.

(b) It should be possible to replace the cell reproducibly in the cell holder. Ingle [3] has tested two simple commercial spectrophotometers and shown that, for both, the most serious loss of precision was due to poor cell positioning. Difficulties may arise through design faults in the cell-holder, weakening or corrosion of the locating springs, or a build-up of dirt in the cell-holder. A cell with chipped corners or which is under-sized may not locate properly.

To test the reproducibility of the alignment of cells, it will be necessary to check first the reproducibility of the instrument itself by repeating the measuring routine several times without the cell in the holder. The reproducibility associated with the cell alignment can then be checked by attaching to or inserting into a test cell a mask with a vertical slit 1 mm wide. The monochromator is turned to 550 nm, the slits of the instrument opened, and visual inspection made to ensure that most of the light passes through the aperture in the cell. The apparent transmission is measured, the cell removed, replaced and measured again. This is repeated several times. The measured transmissions should be within better than 1% of their mean value.

Despite the general historical recognition of cell positioning as one of the most significant factors in absorbance precision, it has not proved possible to design a cell-holder which has eliminated the problem. For this reason, it is recommended that conventional cells be filled and refilled by pipette for quantitative work or that flow-through cells be used if appropriate.

For measurements of the highest precision, it is recommended that a special cell holder is made so that the cell is fixed firmly in position by a screw device. The cell is then clamped in position for the duration of the series of measurements. Figure 3.3 shows a satisfactory design for a cell-holder which locates a Normal 10 mm cell by means of a screw clamp. Note that the chamfer needs to be at an angle at which it will not direct extraneous light into the cell.

(c) The cell-holder should reposition accurately when it is removed from the instrument or, if it is on a slide or a turret, when it is moved from one position to another and back again. This can also be checked using the aperture cell described above.

(d) The positions of a multi-cell-holder should be equivalent. Again, the apparent transmission of the aperture cell should be the same when it is placed in different positions.

3.6 Matching cells with instrument

The response of photocells and photomultipliers is sensitive to the position at which the beam enters the window. This positional sensitivity

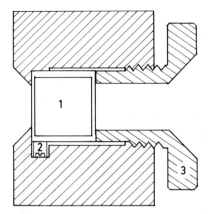

Fig. 3.3 Horizontal section of a special cell-holder with clamping screw used in the Radiation Dosimetry Laboratory, NPL. 1: 10 mm rectangular cell; 2: Leaf-spring ensuring lateral location of the cell; 3: Clamping screw.

depends on the detector design, the quality of the photosensitive coating, the wavelength and the surface of the detector. Vertically mounted side-window photomultipliers are particularly sensitive to horizontal variation in the position of the beam. Care must be taken to ensure that the beam moves as little as possible and in a reproducible manner when the cell is introduced. This means that the cell must have its windows perpendicular to the beam, and the surfaces of the windows must be as parallel as possible. It must be remembered that plastic cells will cause a significant vertical deviation since they are tapered as a result of the moulding process.

The quality of cells and detectors is to some extent complementary: a poor cell can give satisfactory results if the detector has a uniform sensitivity over its surface. If the detector in an instrument is changed, it is advisable to check the performance of cells and cell-holders used for the most accurate measurements.

The more expensive instruments will in general give better performance, as well as offering greater versatility. This may be of particular importance in respect of photometric accuracy and stray-light. However, the performance of even the cheaper instrumentation has improved dramatically as a result of the advance of modern technology over the last 20 years. Since it is improbable that user skill has shown the same degree of improvement, the gap between the sources of error has widened. Some experimental robust cells were fabricated from squared silica tubing. Despite their obvious defective appearance with irregularities and striations, they gave surprisingly good reproducibility of measurements on high-quality instruments, although not as good as with proper optically polished cells, which underlines the point being made in the previous paragraph. It was nevertheless felt that there was no merit in reducing instrument performance by using inferior accessories and the project was abandoned. That message must be even more true today, that it only does justice to good instrumentation to use good quality cells. There are now spectrometers available with a specification of absorbance accuracy of 0.00003 Absorbance Units at 0.1 AU. To obtain the maximum performance, the greatest care needs to be taken over the choice, care and use of cells and instruments.

3.7 Self-calibrating instruments

In response to regulatory pressures and to try to help users to avoid having to perform repetitive calibrations, instrument companies have put many self-calibrating instruments on the market recently. When switched on, the instrument checks its settings against suitable in-built standards. For example, the wavelength scale is checked against the positions of the deuterium lines from the source and appropriate adjustments made if there is a discrepancy. Only when all parameters have been checked and normalized can the instrument be used. It might be thought that this would eliminate the need for the user to bother with standardization and the recommendations in this book. However, careful users and the regulatory authorities have recognized both an inadequacy in the procedure and an underlying problem. The algorithms for calibration, the reliability of the procedures and often even the bases of the calibration procedures are hidden from the user so that it is difficult to be sure for oneself and therefore to take responsibility for the state of calibration of the instrument. It is therefore necessary to perform a confirmatory check. In the case of a non-self-calibrating spectrometer, this can be carried out at intervals commensurate with the user's experience of the stability of the instrument with time. For a self-calibrating spectrometer, it is necessary to re-check the calibration each time it is switched on, in case a subsequently undetectable adjustment has taken place. The implications of these considerations will depend on the area of work and the purpose of the measurements, the accuracy and precision sought, and whether GLP is involved, but users ought to be aware of the situation.

3.8 Diode-array instruments

In the diode array spectrometer, the light beam from the spectrometer passes through the sample before entering the dispersion system. Four different beam geometries are possible: the rays could come to a focus within the cell as shown in Figs 3.1 and 3.2, or they could be parallel, convergent or divergent. In practice, only parallel beam and convergent beam geometries are employed. The focused beam is avoided because of the need to minimize the potential photolytic effect of the unmonochromated beam on the sample. The consequence of these beam geometries is that the beam shape is more nearly square/circular than in photomultiplier instruments.

Most of the considerations of the interaction of cell and beam geometry discussed above apply *mutus mutandi* to the diode array instrument. Parallel or convergent beam systems indeed simplify the correct positioning of Micro cells and long pathlength cells. However, there are additional factors to be taken into account due to the placing of the monochromator after the sample. The first is that any distortion of the beam by the cell which results in a change of angle at which the beam is driven through the monochromator will result in an error in the wavelength scale. The effect is substantial. A cell with a wedge of 3' could cause a wavelength shift of as much as 1–2 nm, so altering intensity ratios within the spectrum as well as causing energy loss at the extremes of wavelength. The second factor involves the equivalent consideration to that discussed under detectors. The entrance slit of the monochromator is in effect a very non-uniform detector, so that any deviation of the beam from its intended path will result in loss of energy due to interception by the slit jaws. Instruments incorporate a number of design features to eliminate these effects. One is to place the cell against the slit so as to minimize the effect of lateral beam shift, but this will not improve any consequence of angular deviation of the beam. Such a situation could be caused by a cell misplaced in the cell-holder or by cells with faces not accurately at right angles to the sides. This implies that it is desirable to always use the highest quality cells in diode array instruments. In convergent beam diode array instruments, it has been shown that interreflection errors from coated transmission standards are larger than in other instrument designs.

3.9 HPLC detectors

A spectrometer system to monitor eluate from a chromatographic column must be designed not to interfere with the optimum column performance. Hence, the spectrometer must match the sample rather than the sample match the spectrometer. The sample from the column is small and flowing, so some compromise in spectrometer performance is inevitable. Rapid-scanning or multi-channel instruments, e.g. diode array spectrometers, are desirable so as to collect the whole spectrum during the passage of the chromatographic peak through the illuminated area. The small sample volume, cylindrical cell geometry and the need to minimize defocusing due to refractive index changes makes it essential that the beam is focused tightly in the sample volume. Chromatographers demand the option of monitoring at wavelengths as low as 190 nm so as to pick up compounds lacking chromophores, so the optical path from source to detector needs to be small so as to avoid purging.

All these factors contribute to low signal-to-noise characteristics of the signal. This and the problem of arranging truly repeat measurements render accurate performance checks more difficult. However, no calibration can be better than the system performance allows. Absorbance repeatability and linearity rather than absorbance accuracy is the key issue. Wavelength accuracy is not a demanding requirement because of the limited resolution inevitable in these systems.

References

1. Threlfall, T.L. (1988) *European Spectroscopy News*, **78**, 8.
2. Commission Internationale de l'Eclairage, Publication 15.
3. Ingle, J.D. (1977) *Anal. Chim. Acta*, **88**, 131.

4 Liquid Absorbance Standards

4.1 Introduction

The advantage of using a solution as an absorbance standard is that the procedure for its measurement closely resembles that for a normal sample. The use of solution standards combines both operator error and cell errors in one set of measurements and, providing that Beer's law is obeyed for the standard concerned, can be used at any absorbance level in the instrument's range. Liquid standards are also ideal for calibrating flow systems, e.g. high-performance liquid chromatography detectors, as they reproduce the cell dimensions precisely [1]. The disadvantages of solution standards are that they do not have the high precision or optical neutrality found with solid filters, and they usually have larger temperature coefficients than solid standards. The lack of optical neutrality means that the precision will vary very markedly with wavelength, and usually only measurements at a maximum or minimum can be recommended. Additionally the precision, especially in standards with sharp bands, will be dependent on the spectral slit width used.

4.2 Standards for the 200–400 nm region

4.2.1 Potassium dichromate

Of all the compounds suggested for the preparation of absorbance standards, potassium dichromate has been the most extensively studied. The choice of solvent conditions is the point most in dispute, for both acidic and alkaline aqueous media have their advantages and disadvantages.

Potassium dichromate in acid solution

The NBS has reported detailed studies on potassium dichromate in acidic media [1–4]. In solution, the following equilibrium processes occur:

$$H_2CrO_4 \overset{K_1}{\rightleftharpoons} H^+ + HCrO_4^- \qquad K_1 = 0.16\ [5]$$
$$HCrO_4^- \overset{K_2}{\rightleftharpoons} H^+ + CrO_4^{2-} \qquad K_2 = 3.2 \times 10^{-7}\ [6]$$
$$3.0 \times 10^{-7}\ [7]$$

$$2HCrO_4^- \overset{K_3}{\rightleftharpoons} Cr_2O_7^{2-} + H_2O \qquad K_3 = 43.7\ [5];\ 35.5\ [8]$$
$$33.0\ [9];\ 32.9\ [2]$$

The dissociation constants given above were all measured at 25°C. At pH = 10, the chromium (VI) present exists as 99.9% CrO_4^{2-}, whereas in weakly acidic solution (approximately pH = 3), the predominant species is $HCrO_4^-$ with less than 0.1% contribution from H_2CrO_4. As indicated above, $HCrO_4^-$ is capable of dimerization, and it is this process that causes difficulty when measuring the absorption spectrum under acid conditions. The amount of dimer formed is dependent on the initial concentration of the salt present: at 1.36×10^{-4} M, 0.9% dimer is formed; while at 6.80×10^{-4} M, 4.2% dimer is present [3].

Figure 4.1 shows the absorption spectrum at a concentration of 1.0×10^{-4} M, while Fig. 4.2 shows that the molar absorptivity of the dimer is greater than that of $HCrO_4^-$ at most wavelengths, and so it is the pre-

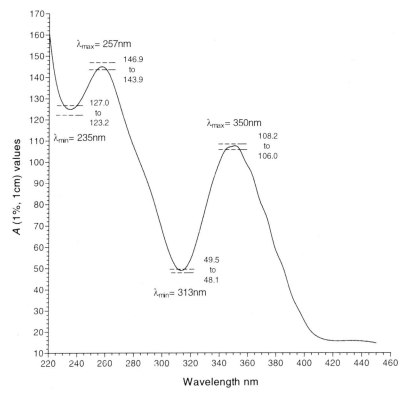

Fig. 4.1 Potassium dichromate in 0.005 M sulphuric acid solution (61.2 mg per litre) at 20°C, 10 mm pathlength and 1 nm SBW. These data are displayed as A (1%, 1 cm) values corrected for solvent blank. Jasco V560 spectrometer, scan rate 40 nm min^{-1}, medium detector response and data pitch 0.1 nm.

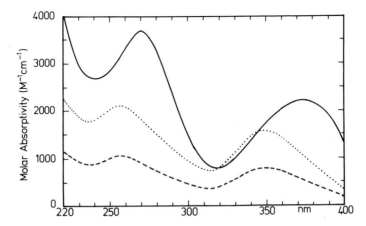

Fig. 4.2 Computed absorption spectra for the chromate (- - - -) and dichromate (———) ions in perchloric acid, pH = 3.0. The curves were derived by extrapolation from the spectra of dichromate solutions of various concentrations. The spectrum of the chromate ion multiplied by 2 is also shown (....). At the isosbestic points of this and the dichromate curve the solution should obey Beer's law. Redrawn from Ref. [3].

sence of dimer that causes the observed deviations from Beer's law [2, 10]. This is the main drawback to the use of acid solutions as a standard.

The choice between the use of sulphuric acid or perchloric acid as solvent is a less important issue. The NBS points out that 0.005 M H_2SO_4 has two disadvantages compared with perchloric acid at pH = 3: sulphuric acid has a greater ionic strength and hence greater salt effects; and there is the possibility of forming mixed chromium (VI)–sulphate complexes [3].

Since the values of the dissociation constants K_n are dependent upon the acidity of the solution [8], when using 0.005 M H_2SO_4 at pH = 2 there is an increased possibility of deviations from Beer's law. At this pH, it is possible for H_2CrO_4 to be present as well, and this causes even more uncertainty in the observed absorbance values. However, several workers prefer sulphuric acid, as it is safer to handle and more readily available. In general, it is accepted that a pH of 3 is preferable to a pH of 2, but it should not be greater than 3. Whichever acid is used, it is essential that the nature and concentration of the solvent is quoted, that the pH of the solution is accurately measured and quoted, and the concentration of the potassium dichromate also given.

The molar absorptivity of potassium dichromate in acid solution is sensitive to temperature. At 313 nm the temperature coefficient is 0.02% per °C, while at 235, 257 and 350 nm it is –0.05% per °C over the range 20–30°C [2]. The photochemical stability of the solution is high, there being no significant changes in absorbance on prolonged exposure to a 250 W high-pressure mercury arc lamp [11].

Much work was carried out between 1950 and 1970 to determine the molar absorptivity of potassium dichromate [3, 10, 12–21]. A survey of the data led the UV Spectrometry Group to recommend two concentrations of potassium dichromate for use as calibration standards. Solution A contained 50 ± 0.5 mg in 1 litre of 0.005 M sulphuric acid and was intended for the absorbance range 0.2–0.7 A. Solution B contained 100 ± 1 mg in 1 litre of 0.005 M sulphuric acid and was intended for the absorbance range 0.4–1.4 A. The Group recommended the absorbances shown in Table 4.1.

Table 4.1 Recommended absorbance values for potassium dichromate in 0.005 M sulphuric acid solutions.

Wavelength (nm)	Molar Absorptivity	A (1%, 1 cm)	Absorbance Solution A	Absorbance Solution B
235 (min)	3683.5	125.2	0.626 ± 0.009	1.251 ± 0.019
257 (max)	4277.8	145.4	0.727 ± 0.007	1.454 ± 0.015
313 (min)	1435.7	48.8	0.244 ± 0.004	0.488 ± 0.007
350 (max)	3153.9	107.2	0.536 ± 0.005	1.071 ± 0.011

Many of the trials that led to the values in Table 4.1 were carried out on older instruments that were, perhaps, not of the standard of today's instruments. Encouragingly, the values in Table 4.1 were confirmed by a study on a modern double monchromator instrument which looked at ten concentrations of potassium dichromate [22]. This study found that 350 nm was the best wavelength for the use of potassium dichromate as a transfer standard. Values for the molar absorptivity at 10 nm intervals over the range 200–500 nm were also reported, and several wavelengths identified as useful. The data are reproduced in Table 4.2.

A collaborative trial involving 16 instruments including diode array instruments as well as double and single beam instruments was carried out by the European Pharmacopoeia Commission. This study involved the

Table 4.2 Molar absorptivity for potassium dichromate in 0.005 M sulphuric acid at recommended wavelengths [22].

Wavelength (nm)	Mean molar absorptivity (for $HCrO_4^-$)	Absorbance (50 mg/l)
250	2064.6 ± 18.6	0.702 ± 0.003
313	714.9 ± 5.9	0.243 ± 0.001
320	818.8 ± 8.6	0.278 ± 0.001
340	1488.4 ± 9.5	0.506 ± 0.001
345	1567.9 ± 9.6	0.533 ± 0.002
350	1579.4 ± 6.4	0.537 ± 0.001

measurement of two batches of potassium dichromate, one common to all instruments and one supplied by each laboratory [23]. The results are shown in Table 4.3. The values obtained at 350 nm are closest to the values in Tables 4.1 and 4.2, and confirm this as the most reliable wavelength.

It has been suggested that one of the isosbestic wavelengths of the dimer and $HCrO_4^-$ spectra, e.g. 345 nm, could be used for standard absorbance measurements, since there should be no deviation from Beer's law [10] and this wavelength has been recommended [22], see Table 4.2.

Potassium dichromate in alkaline solution

Alkaline potassium dichromate has not been used extensively in collaborative trials. However, Haupt of the NBS has made detailed studies

Table 4.3 Specific absorbances of potassium dichromate in 0.005 M sulphuric acid from a collaborative trial [23].

Instrument type	Wavelength (nm)							
	235		257		313		350	
Double-beam	125.0	125.7	145.0	145.3	48.9	49.2	107.8	107.7
Double-beam	124.3	124.0	144.2	144.2	48.5	49.8	107.2	107.9
Double-beam	123.3		143.3		47.9		106.5	
Double-beam	125.5	124.8	144.2	143.9	48.5	49.1	107.6	107.0
Single-beam	125.5	124.8	145.4	145.0	48.8	48.6	107.8	107.5
Diode array	124.2	124.3	144.3	144.4	48.6	48.7	106.7	106.8
Diode array	123.3	123.7	143.5	144.0	48.2	48.4	106.1	106.4
Diode array	124.1	124.1	144.3	144.4	48.2	48.3	106.7	106.6
Double-beam	125.5	126.2	(145.8)*	145.9	49.1	49.2	108.1	107.9
Double-beam	124.6		144.7		48.5		107.4	
Single-beam	124.4	123.7	144.4	143.7	48.3	48.2	107.1	107.2
Double-beam	123.8	123.9	144.2	144.3	48.3	48.3	107.1	107.3
Double-beam	124.6	124.6	145.1	143.2	49.0	48.7	107.4	107.7
Double-beam	(127.8)*	128.0	(147.4)*	147.7	49.6	49.8	106.5	106.8
Diode array	125.5	123.6	145.5	143.6	49.2	48.1	107.9	106.5
Grand mean	124.5		144.5		48.7		107.2	
Standard deviation	1.0		0.9		0.5		0.6	
Corresponding aborbance of mean at a concentration of 0.05 g/l	0.623		0.723		0.243		0.536	

*Values excluded by the reference authors. The first column at each wavelength represents the same batch of potassium dichromate

and recommends that, as potassium dichromate is commercially available in a purer state than potassium chromate, the former should be used to make the solutions, even though the solutions contain effectively only potassium chromate and are identical with solutions of potassium chromate in 0.05 M potassium hydroxide [24]. Figure 4.3 gives a typical absorption spectrum.

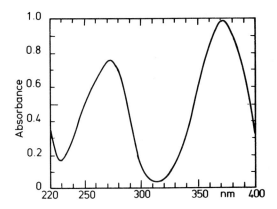

Fig. 4.3 Absorption spectrum of 1.0×10^{-4} M potassium dichromate in 0.05 N potassium hydroxide measured in a 10 mm cell. Temperature and pH not specified. Redrawn from Ref. [3].

Haupt proposes that the standard solution should be prepared from 0.0300 g of potassium dichromate in 1 l of 0.05 M aqueous KOH, which will give a solution containing 0.0400 g l^{-1} potassium chromate. The solution was found to be spectroscopically stable above 260 nm over a period of 6 years. Changes at wavelengths below 260 nm were observed after 6 months, and so care must be taken if measurements are to be made at short wavelengths. The solutions should be stored in glass bottles – there will be a tendency for 'flaking' to occur, but if this is allowed to settle, it should not affect the absorbance values. The solutions obey Beer's law, and the temperature coefficient at 373 nm is 0.09% per °C, and at 274 nm is 0.06% per °C over the temperature range 17–37°C [2]. The results of studies at the NBS [2] are shown in Table 4.4, they are broadly in line with earlier studies [21, 24–31]

Potassium chromate in disodium hydrogen phosphate solution

It has been suggested that 0.05 M Na$_2$HPO$_4$ could be used in place of KOH to overcome the difficulties of handling strongly alkaline solutions [2, 32]. Deviations from Beer's law do not exceed 0.10%. Values for the molar absorptivities at two wavelengths are given in Table 4.5.

Table 4.4 Apparent molar absorptivity of potassium chromate in 0.05 M KOH at 25°C.

K_2CrO_4 concn (M) Wavelength	Molar absorptivity	
	274 nm	373 nm
7×10^{-5}	3705	4830
14×10^{-5}	3698	4824
21×10^{-5}	3691	4814

Table 4.5 Apparent molar absorptivity of potassium chromate in 0.05 M Na_2HPO_4 at 25°C [2].

K_2CrO_4 concn (M) Wavelength	Molar absorptivity	
	274 nm	373 nm
7×10^{-5}	3703	4827
14×10^{-5}	3697	4820
21×10^{-5}	3692	4813

Comparison of dichromate systems

Dichromate in acid solution has become the definitive liquid absorption standard. The acid spectrum has a better arrangement of absorbance maxima and minima than the alkaline solution. The effect of pH upon the equilibria under acid conditions makes it essential that stringent standard conditions are laid down for both pH and potassium dichromate concentration.

Although the stability and linearity of the alkaline solution has been established, the majority of work has been carried out on the more temperamental acidic solutions. The corrosive properties of KOH seem to be disliked by most workers.

4.2.2 Potassium hydrogen phthalate

Potassium hydrogen phthalate is obtainable from the NIST as a high-purity standard for acidimetry. As shown in Fig. 4.4, the spectrum is dependent upon pH, and Burke *et al.* chose 1% aqueous perchloric acid, pH = 1.3, as a solvent [2]. Under these conditions, the solution contains 97% phthalic acid and 3% hydrogen phthalate ions. Their molar absorptivity values at two concentrations are given in Table 4.6.

The temperature coefficient is −0.05% per °C at 275.5 nm, and +0.05% per °C at 262 nm over the range 17–37°C. Weak fluorescence has been observed at 350 nm when exciting at 280 nm, but the full effects of this have not yet been estimated.

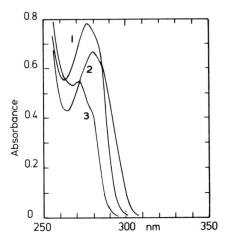

Fig. 4.4 Absorption spectra of aqueous solutions of potassium hydrogen phthalate measured in a 10 mm cell: (1) 0.122 g l^{-1} in 0.1 N perchloric acid; (2) 0.103 g l^{-1} in water; (3) 0.141 g l^{-1} in 1.0 N potassium hydroxide. Redrawn from Ref. [2].

4.2.3 Nicotinic acid

Nicotinic acid was proposed by Milazzo et al. as a suitable absorbance standard for shorter wavelengths [12]. The compound is available in a highly pure state and, if 0.1 M hydrochloric acid is used as solvent, the amino group will be fully protonated. The spectrum is shown in Fig. 4.5, and has a shoulder at 210 nm and a peak at approximately 265 nm. The spectrum is not ideal for an absolute standard. The shoulder at 210 nm may be susceptible to stray-light, and the peak at 265 nm is quite sharp and will be influenced by spectral slit width.

Nicotinic acid is probably best used to check the linearity of the system. A solution containing 0.023 g L^{-1} will give an absorbance of about 1 A. The concentration can be varied around this value to assess the linearity of the spectrometer at 210 nm using the shoulder and at 265 nm using the maximum [Burgess, C., personal communication].

Table 4.6 Molar absorptivities for potassium hydrogen phthalate in 1% perchloric acid, pH = 1.3, at 25°C [2].

Concentration (g/l)	Molar absorptivity	
	262 nm (min)	275.5 nm (max)
0.034	917.8	1293.1
0.142	916.7	1290.5

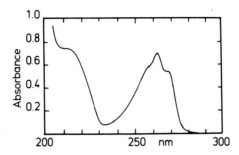

Fig. 4.5 Absorption spectrum of 0.164 g kg^{-1} nicotinic acid in 0.1 N hydrochloric acid measured in a 10 mm cell [26].

4.2.4 Potassium nitrate

The spectrum of potassium nitrate has a single maximum at 302 nm as shown in Fig. 4.6. Collaborative trials were carried out on potassium nitrate [1–3, 33–40], but the results of the trials were so variable that it was not recommended as a standard. However, recently potassium nitrate has been recommended as a standard [41]. A solution of 14.20% w/v in distilled water is prepared and diluted to give concentrations of 1.065%, 0.710% and 0.355% w/v. A value for the A 1% 1 cm of 0.705 is assumed and the corresponding absorbances are 0.751, 0.500 and 0.250. A limit of ±0.010 A is recommended.

4.3 Standards for the 400–800 nm region

4.3.1 Cobalt ammonium sulphate

This compound has been utilized in several surveys [13, 42]. An aqueous acid solution obeys Beer's law over a considerable absorbance range, and

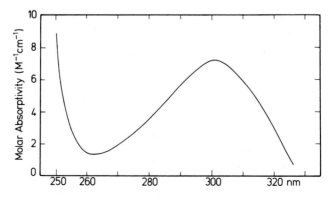

Fig. 4.6 Absorption spectrum of an aqueous solution of potassium nitrate.

so by varying the concentration a wide photometric range can be covered [43]. A typical absorption spectrum is shown in Fig. 4.7. The absorbance of the solution at wavelengths up to 680 nm increases with temperature [44].

The spectrum shows a relatively sharp maximum and the absorbance values obtained will be influenced by the spectral slit width of the instrument [13]. Table 4.7 summarizes the data available at the maximum. Concentrations of $CoSO_4(NH_4)_2.6H_2O$ varied from 14.481 g to 21.1 g in 1 l of 1% aqueous sulphuric acid (specific gravity = 1.835) [16,43,45,46]. The data from wide band instruments, spectral slit width > 10 nm, were so variable that cobalt ammonium sulphate cannot be recommended as a standard for such instruments [13].

Readings from two wavelengths either side of the maximum are summarized in Table 4.8 and indicate that the data become more variable. The

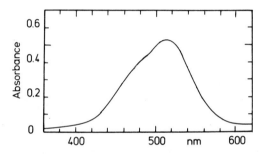

Fig. 4.7 Absorption spectrum of cobalt ammonium sulphate in 1% sulphuric acid, concentration unspecified [42].

Table 4.7 Molar absorptivity and corresponding absorbance for cobalt ammonium sulphate at 510 nm.

Reference	Concentration (M)	Molar absorptivity	Absorbance	Absorbance at 0.05 M
16	0.0367	4.74	0.174	0.237
45	0.0504	4.75	0.240	0.238
16	0.0735	4.71	0.346	0.235
13	0.05	4.84 ± 0.22	0.242 ± 0.017	0.242
13	0.1002	4.85 ± 0.17	0.486 ± 0.011	0.243
46	0.0735	4.71	0.346 ± 0.005	0.235
Mean		4.77		
Standard deviation		0.063		
Coefficient of variation (%)		1.3		

Table 4.8 Molar absorptivities for cobalt ammonium sulphate at two wavelengths away from the maximum.

Reference	Molar absorptivity	
	450 nm	550 nm
45	2.11	2.12
16	2.04	2.07
46	2.07	1.99
Mean	2.07	2.06
Standard deviation	0.034	0.066
Coefficient of variation (%)	1.7	3.2

use of 450 nm is perhaps acceptable, but the use of 550 nm cannot be recommended.

4.3.2 Other compounds

The use of cobalt and nickel ions has been examined by the NBS [2], and whilst the solutions have useful spectra, the preparation is difficult and they are not recommended as standards. Solutions of cobalt and nickel are available from NIST with certified absorbance values, see Section 4.6.1.

Copper sulphate has been examined as a standard but the absorption below 600 nm is so low as to preclude its use [43, 45].

4.4 Standards for the entire 200–800 nm region

4.4.1 Green food colourings

Langdale's Sap Green, consisting of Green S and Tartrazine has been suggested by Burgess as a standard for the whole of the UV visible region [47]. The standard is prepared by diluting 5.0 ml of the colouring to 500 ml with water, and 20 ml of this solution is diluted to 100 ml with water. The absorbance was found to be independent of pH over the range 2–6.5, and the values measured at 238, 425 and 634 nm were constant over at least 30 days. A further, unpublished, survey conducted by Burgess used a more concentrated solution made by diluting 5 ml of the Langdale's Sap Green stock solution to 2 l with water. The solution remained stable for over 1 year, and the only deterioration noted was a slight bacterial growth that did not affect the absorbance values.

These dye solutions can only be used as linearity checks or transfer standards because of the difficulty in obtaining the pure components. They have the advantage of being safe, inexpensive, very stable and easy to use.

4.5 Choice of standards

4.5.1 Choice of standards for the 200–400 nm region

Of the standards reviewed in this section, potassium dichromate has been studied the most and, despite some unsatisfactory properties, potassium dichromate in 0.005 M sulphuric acid seems to be the best standard to use. The use of perchloric acid is preferred by some workers. Great care must be taken to control the pH of the solution.

Alkaline solutions have been shown to be stable, although 'flaking' may occur. These solutions obey Beer's law and the absorbance values are less sensitive to pH than the acid solutions. However, acid dichromate has become the recommended standard for this region.

Alternative compounds to acid dichromate have been examined, but none has become established as recommended standards. The spectrum of potassium hydrogen phthalate is very pH dependent, and little work has been done on these solutions beyond that reported by the NBS. Potassium nitrate appears to be an ideal standard, with a single broad peak of low molar absorptivity. However, the uncertainty over the molar absorptivity values makes this an unsatisfactory standard. Nictotinic acid contains a shoulder at 210 nm which may be useful at low wavelengths, but stray-light may be a problem.

4.5.2 Choice of standards for the 400–800 nm region

The best standard in this group would appear to be acidic cobalt ammonium sulphate. Although recommended values are given away from the maximum [46], little work has been published to justify the use of such wavelengths, and the use of 510 nm is recommended. Availability can be a problem as few companies stock the chemical in a pure form.

Solutions of cobalt(II) and nickel(II) ions have been examined by the NBS, both as the sulphate and the perchlorate [2]. The solutions can be purchased from the NIST, see Section 4.6.1. In the UK, the Laboratory of the Government Chemist supply similar solutions, see Section 4.6.4. The preparation of the solutions is not easy, and commercial solutions are relatively expensive.

4.5.3 Choice of a standard for the 200–800 nm region

There is no simple absorbance standard that will satisfactorily cover the whole UV–visible range, and it is necessary to have separate solutions for the UV and visible regions.

4.6 Commercially available standards

There are now several sources of solutions of certified absorbance that can be used as standards in the UV and visible regions. A selection of these is given below.

4.6.1 SRM 931: Cobalt and nickel perchlorates

Standard Reference Material 931, National Institute of Standards and Technology, Gaitersburg, MD 20899–001, USA. This was the first absorbance standard to be issued by NIST (or NBS as it was then called) [2]. Solutions of three different concentrations and a blank are supplied in sealed ampoules, together with a certificate of absorbance at four wavelengths. The solutions are prepared by dissolving high-purity cobalt and nickel in a mixture of nitric and perchloric acids. The weights are chosen so that the maxima are of approximately the same height. A typical spectrum is given in Fig. 4.8. The concentration of nitrate ion is reduced to that of the metallic ions by fuming, giving a solution of pH = 1.

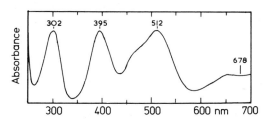

Fig. 4.8 Absorption spectrum of NBS Standard Reference Material 931. A set of three solutions is supplied, of maximum absorbance approximately 0.3, 0.6 and 0.9 [2].

The absorbance is given at the maxima at 302, 395, 512 and the plateau at 678 nm. The nominal absorbances are 0.3, 0.6 and 0.9 at wavelengths of 302, 395 and 6512 nm, respectively. At a wavelength of 678 nm the nominal absorbances of the three solutions are 0.1, 0.2 and 0.3, respectively. The uncertainties in the values are given at the 95% confidence level and include a systematic error of 0.5%. The absorbance is certified at 25°C, but the corresponding value at any temperature between 17 and 27°C can be calculated from the following formula:

$$A_t = A_{25} + (t - 25).C_A$$

where A_t and A_{25} are the absorbances at the temperature of observation and 25°C, and C_A is the temperature coefficient, i.e. −0.0014 at 302 nm; +0.0014 at 395°C; +0.0018 at 512 nm and +0.0014 at 678 nm.

Further details are available on the NIST website: *http://ts.nist.gov/srm*

4.6.2 Starna potassium dichromate

Contact: Starna, 33 Station Road, Chadwell Heath, Romford, Essex RM6 4BL, UK. Potassium dichromate standards are supplied in six sealed 10 mm cells with nominal absorbances of 0, 0.2, 0.4, 0.6, 0.8 and 1.2 at 350 nm. The temperature coefficient is less than 0.1% per °C over the range 20–30°C, and the samples are claimed to be stable indefinitely. The manufacturers recommend that these cells should be used as secondary standards only.

4.6.3 SRM 935: solid potassium dichromate

Solid potassium dichromate is available from the Office of Standard Reference Materials – see Section 4.6.1 for their address and website. SRM 935 is crystalline potassium dichromate of certified purity which is supplied with extensive details of the preparation of solutions, measurement of absorbance and the expected molar absorptivity values. It is described in detail by Burke and Mavrodineanu [4].

4.6.4 Laboratory of the Government Chemist

Contact: Laboratory of the Government Chemist, Queens Rd, Teddington, Middlesex TW11 LY, UK. The Laboratory of the Government Chemist supply a solution of sodium, cobalt and nickel salts in dilute perchloric acid. A set of four sealed silica cuvettes containing three solutions (of varying concentration) and one blank is supplied. The solutions provide a range of certified absorbance values when measured against the blank supplied. The solutions are certified at four wavelengths (299, 394, 512 and 719 nm) at a bandwidth of 1 nm. The temperature of the solutions needs to be controlled when taking measurements.

4.7 Conclusions

The majority of solutions mentioned in this chapter can be quickly prepared from readily available compounds and allow a very simple check on the spectrophotometer.

Most of these solutions are stable over long periods, and so a reasonable quantity can be prepared at one time. It is therefore recommended that solution standards are prepared in the laboratory, using the purest available compounds, in the case of potassium dichromate SRM 935 is recommended. The commercially available solutions are expensive and offer no particular advantage, unless time or the necessary expertise are at a premium. If traceable liquid standards are required in small quantities, however, it may be more cost-effective to purchase them.

Procedures for the preparation and measurement of a dichromate standard are given in Chapter 9.

References

1. Menis, O. and Schultz, J.T. (1970) NBS Technical Note 544.
2. Burke, R.W., Deardorff, E.R. and Menis, O. (1972) *J. Res. NBS*, **76A**, 469.
3. Menis, O. and Schultz, J.I. (1971) NBS Technical Note 584.
4. Burke, R.W. and Mavrodineanu, R. (1977) NBS Special Publication No. 260–54.
5. Ringbom, A. (1963) *Complexation in Analytical Chemistry*. Wiley, New York.
6. Neuss, J.D. and Rieman, W. (1934) *J. Amer. Chem. Soc.*, **56**, 2238.
7. Howard, J.R., Nair, V.S.K. and Nancollas, G.H. (1958) *Trans. Faraday Soc.*, **54**, 1034.
8. Tong, J. and King, E.L. (1953), *J. Amer. Chem. Soc.*, **75**, 6180.
9. Davis, W.G. and Prue, J.E. (1955) *Trans. Faraday Soc.*, **51**, 1045.
10. Burke, R.W. and Mavrodineanu, R. (1976) *J. Res. Nat. Bur. Std.*, **80A**, 631.
11. West, M.A. and Kemp, D.R. (1976) *Int. Lab.*, May/June, 27.
12. Milazzo, G., Palumbo-Doretti, S.C.M. and Violante, N. (1977) *Anal. Chem.*, **49**, 711.
13. Vanderlinde, R.E., Richards, A.H. and Kowalski, P. (1975) *Clin. Chem. Acta*, **61**, 39.
14. Gridgeman, N.T. (1951) *Photoelec. Spec. Grp. Bull.*, **4**, 67.
15. Ketelaar, J.A.A., Fahrenfort, J. Haas, C. and Brinkman, G.A. (1955) *Photoelec. Spec. Grp. Bull.*, **8**, 176.
16. Rand, R.N. (1969) *Clin. Chem.*, **15**, 839.
17. Bouche, R. and Molle, L. (1975) *J. Pharm. Belg.*, **30**, 578.
18. Vandenbelt, J.M. (1960) *J. Opt. Soc. Amer.*, **50**, 24.
19. Wylie, E., unpublished observations, quoted in Ref. [14].
20. Inman, S.R., unpublished observations, quoted in Ref. [14].
21. Morton, R.A. (1951) *Photoelec. Spec. Grp. Bull.*, **4**, 65.
22. Gil, M. Escolar, D. Iza, N. and Montero, J. (1986) *Applied Spectroscopy* **40**(8), 1156–1161.
23. *Pharmeuropa* (1993) **5**(2), 156–165.
24. Haupt, G.W. (1952) *J. Opt. Soc. Amer.*, **42**, 442.
25. Vandebelt, J.M., Forsyth, J. and Garret, A. (1945) *Ind. Engrg Chem. Anal., (Ed.)* **17**, 235.
26. Brode, W.R., Gould, J.H. Whitney, J.E. and Wyman, C.M. (1953) *J. Opt. Soc. Amer.*, **43**, 862.
27. Vandenbelt, J.M. and Spurlock, C.H. (1952) *J. Opt. Soc., Amer.*, **43**, 862.
28. Hogness, T.R. and Zscheile, F.P. (1947) Unpublished observations.

29. Rosslet, G. (1926) *Chem. Ber.*, **59**, 2606.
30. Hogness, T.R., Zscheile, F.P. and Sidwell, A.E. (1937) *J. Phys. Chem.*, **41**, 379.
31. *Collaborative Spectrophotometric Study* (1940) US Pharmacopoeia.
32. Johnson, E.A. (1967) *Photoelec. Spec. Grp. Bull.*, **17**, 505.
33. Von Halband, H. and Eisenbrand, J. (1928) *Z. Phys. Chem.*, **112**, 620.
34. Everett, A.J., Unpublished observations.
35. Edisbury, J.R. (1949) *Photoelec. Spec. Grp. Bull.*, **1**, 10.
36. Von Halban, H. and Ebert, L. (1924) *Z. Phys. Chem.*, **112**, 329.
37. Scheibe, G. (1926) *Chem. Ber.*, **59**, 2606.
38. Baly, E.C.C., Morton, R.A. and Riding, R.W. (1927) *Proc. Roy. Soc.*, **113A**, 709.
39. Ley, H. and Volbert, F. (1927) *Z. Phys. Chem.*, **130**, 308.
40. Ewing, D.T., Vandenbelt, J.M., Emmett, A.D. and Bird, O.D. (1940) *Ind. Engrg Chem. Anal. Ed.*, **12**, 639.
41. Australian Code for Therapeutic Goods – Guidelines for Laboratory Instrumentation, November 1991, p. 15.
42. Rand, R.N. (1973) NBS Special Publication no. 378, 125.
43. Gibson, K.S. (1949) NBS Circular 484.
44. Davies, R. and Gibson, K.S. (1931) NBS Miscellaneous Publication M114.
45. Kortum, G. (1955) *Kolorimetrie – Photometrie und Spektrometrie*, 3rd edn. Springer, Berlin.
46. *Pharmeuropa* (1993) **5**(3), 213–216.
47. Burgess, C. (1977) *UV Spec. Grp. Bull.*, **5**, 77.

5 Solid Absorbance Standards

5.1 Introduction

Many methods of checking the absorbance accuracy of spectrophotometers that do not use solutions have been developed over the years. It is the purpose of this chapter to highlight the major papers dealing with these methods and, in particular, to discuss those methods most suited to routine laboratory use.

It is relevant at this point to consider what properties an absorbance standard should possess. Ideally it should:

(a) be convenient to use and should simulate the normal use of the instrument;
(b) have an absorbance independent of wavelength setting;
(c) be unaffected by stray-light;
(d) be non-fluorescent;
(e) show little change in its optical properties with temperature;
(f) not be changed by exposure to normal atmospheres and light; and
(g) be easy to construct and calibrate.

It will be readily appreciated that few materials or methods satisfy all of these requirements. In particular, (g) is often difficult to satisfy and rules out many techniques for all but the national standards laboratories or the most ardent spectroscopist.

5.2 Glass filters – a historical perspective

5.2.1 The NBS glass filters

A series of four glass filters was described in a series of papers by NBS staff [1–4] in the period 1934–1955. Typical transmittance curves are shown in Fig. 5.1. Copeland *et al.* [5] studied a Carbon Yellow filter from this series over a period of 6 years. He concluded that the filter was stable, with a day-to-day variability of ± 0.002 absorbance units. The filters were discussed further by Rand [6].

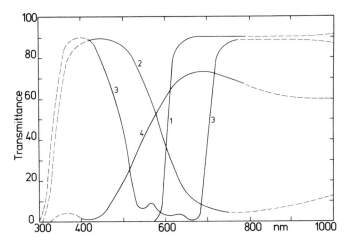

Fig. 5.1 Transmittance spectra for NBS glass filters: 1: Selenium Orange; 2: Copper Green; 3: Cobalt Blue; 4: Carbon Yellow, from Ref. [1].

Criticism of these filters centred around the fact that the values supplied with the filters were certified only to ±3% (0.001–0.0104 depending on the absolute absorbance). The filters all show extensive dependence of absorbance on wavelength setting (at 540 nm the Carbon Yellow curve is rising at a rate of 0.0083 A/nm) and so even minute wavelength setting errors easily resulted in misleading information.

5.2.2 Chance ON10 filters

In the 1960s, attention turned to Chance ON10 as a material. Its use has been discussed by Slavin [7], and Porro and Morse [8]. Its absorbance is shown in Fig. 5.2. A particular sample was measured by Slavin 24 times in six different orientations in the beam, and an absorbance of 1.947 with a standard deviation of 0.004 was achieved. To achieve these figures, the material was polished so that the parallelism was within a few micrometres, and the thickness determined to within 1/2 μm using gauge blocks.

5.2.3 Schott NG-4 glass

Schott NG-4 glass has now become the usual material from which glass absorbance standards are prepared. It is the material used by the NIST in America for SRM 930 and also by the NPL in Britain. The absorbance of NG-4 is shown in Fig. 5.3. The NIST filters are described in a number of papers brought together in a special NIST publication [9] and comprise a

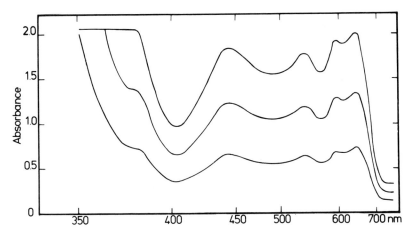

Fig. 5.2 Absorption spectra of Chance ON10 filters of various thicknesses, from Ref. [7].

set of three glass filters to give nominal transmittance of 10, 20 and 30% (1.0, 0.7 and 0.5 A). Filters are also available from the NPL and summarized in refs [11, 12]. These were first described by Sharpe [13] and Popplewell [14, 15], following a cooperative project between NPL and Pye Unicam, and comprise a set of nine filters covering nominal absorbances of 0, 0.14, 0.25, 0.5, 1.0, 1.5, 1.9, 2.5 and 3 A. The filters are manufactured to the most exacting standards and are produced with the filter faces parallel to within 0.5 min of the arc and the faces flat to within two fringes/cm of the mercury green light. They are housed in specially designed stress-free mountings, as shown in Fig. 5.4. It has been pointed out by NIST [16] that the parallelism and flatness requirements are especially important when the filters are used in diode array instruments, as any slight beam distortion due to filter defects will result in the light beam being detected by a different part of the array detector.

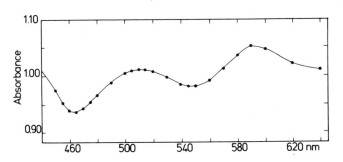

Fig. 5.3 Absorption spectrum of a Schott NG-4 glass filter, from Ref. [13].

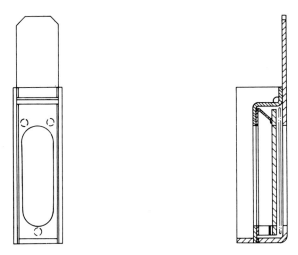

Fig. 5.4 A stress-free mounting for glass filters, from Ref. [13].

5.2.4 Stability and ageing effects

Several workers have investigated the temperature stability of the Schott NG series of glasses – Sharpe [13], Mavrodineanu and Baldwin [17], and Verrill [12]. Mavrodineanu concluded that the variation in transmittance with temperature is insignificant at the 0.95 confidence level. For temperature changes of up to $\pm 2°C$, the variation is minimal and averages less than 0.2% of the measured transmittance values.

The NBS also investigated the stability of the glasses when exposed to a high-intensity source, and concluded that they possess an acceptable stability. Similar tests on other neutral glasses indicated that these were less stable by a factor of 4 compared with the Schott NG-4 glass.

Initial investigations both by the NBS and by Pye Unicam indicated that the time stability of Schott NG glass was very good. The NIST special publication quotes data gathered over a 4 year period which shows a very slight increase in transmittance as would be expected from blooming of the surface. Sets calibrated by NPL for Unicam over the last 20 years confirm this observation.

5.2.5 Reflection correction

Back reflections between the filter faces and other elements in a spectrophotometer (e.g. sample compartment windows) can affect the calibration obtained when using a set of neutral filters. A novel method of overcoming this was described by Sharpe [13], who used a clear glass filter to zero the instrument. This filter compensates for the effect of reflections from the surfaces of the test filter, for these are also present when zeroing the instrument.

A rigorous treatment of the numerical correction required is given by Mielenz and Mavrodineanu [18]. However, this correction has to be obtained from measurements on tilted samples in polarized light and is probably beyond the scope of most laboratories.

5.2.6 Measurement uncertainties

Many factors contribute to the uncertainty levels quoted by suppliers of filters. At the present time, NIST tend to quote a slightly larger uncertainty than NPL as they allow a greater value for filter inhomogeneity. NPL issue filters up to 3 A, and NIST to 2 A.

5.2.7 Traceability

Many of the regulatory bodies now require evidence of traceability of standards used to check spectrometers. This means the standards should be supplied by a National Standards Laboratory (e.g. NPL or NIST) or by a laboratory holding the necessary accreditation. Within the UK, this is provided by the NAMAS scheme, operated by UKAS (United Kingdom Accreditation Service). To be approved under this scheme, a laboratory has got to show evidence of control of the process, with a proper quality system in place and a full understanding of the uncertainties of the process. In the USA, there are moves which will lead to the setting up of suppliers approved to supply NTRM^{TM} filters (NTRM = NIST traceable reference materials).

5.2.8 Other glasses

A comprehensive review of some 800 coloured glasses has been published by Dobrowolski et al. [19]. This lists a number of neutral glasses which could potentially be used as standards, but little has been published on the various parameters (e.g. stability) which are of paramount importance in selecting a filter for use as a standard.

5.3 The use of polarizers to determine photometric linearity

Polarizers have not found wide favour as standards for a variety of reasons, but some very detailed studies of their use has been carried out by Bennett [20], and Mielenz and Eckerle [21]. Bennett points out that three polarizers are required if effects due to the polarization of the optical system of the spectrophotometer are to be avoided. Three polarizers are arranged as shown in Fig. 5.5 with the outer two fixed, and the centre one variable. Absorbance ranges of 0–4.0 are readily achievable by rotation of the centre polarizer. With sheet polarizers the device has a useful

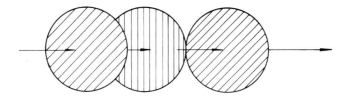

Fig. 5.5 The principle of the three-polarizer attenuator. The outer filters are fixed with their planes of polarization parallel and the centre one is rotated to vary the transmission.

operating range of 375–700 nm. Linearity can be determined to better than 0.1%. Bennett points out the main factors limiting the accuracy of a three-polarizer transmittance standard to be the extinction ratio of the polarizers, the birefringence of the polarizers, the accuracy with which the outer polarizers can be aligned, and the accuracy with which the middle polarizer can be set. Mielenz and Eckerle take the technique further by using quarter-or half-wave plate attenuators in place of the middle polarizer, thus eliminating the problem of unknown birefringence of the middle plate. However, to achieve accuracies of the order of 0.01% in transmittance, the angular setting of the rotating element must be made to within approximately 0.4 min of the arc. The device is not suitable for routine use.

5.4 Metal screens

Metal screens, in the form either of woven wire screens or chemically etched thin metal foils, have been investigated by a number of workers [22–26]. Heidt and Bosley [22] reported the use of woven wire screens in 1953 for checking a Cary spectrophotometer, and in 1961 Newman and Binder [23] reported on the use of a series of screens to provide attenuators with an overall range of six orders of magnitude. Interest in the use of attenuators of this type for the calibration of spectrophotometers revived in the early 1960s, with the search for a standard that was reasonably neutral throughout the 200–800 nm wavelength range.

Vanderbelt [24] and Slavin [25] studied screens formed by chemically etching thin metal plates. The holes in these screens are conical, of uniform size and spacing, with smooth edges. It is always important to present the screen to the light beam in the same orientation, and Vanderbelt recommended that the smaller diameter end of the pores should always face the light beam. Slavin reported variations of greater than 0.03 absorbance units if the same portion of the screen was used in different instruments of the same design, or if the same instrument was used with a different source and detector. He concluded that they were not adequate as routine standards, but could be used for expanding the photometric scale in the region near zero transmittance by placing an appropriate screen in the reference beam of the spectrophotometer. Screens are currently used for this purpose and supplied by a number of manufacturers. Typical absorbance curves for a set supplied by Unicam are shown in Fig. 5.6.

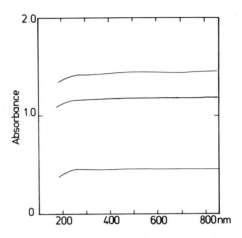

Fig. 5.6 Three typical absorbance spectra for metal screens.

5.5 Sector discs

One approach to the problem of accurately varying the amount of radiation falling on the detector is to use an optical chopper to attenuate the beam, rather than an absorbing filter, e.g. a rotating disc with apertures of known size in it. Hence, the length of time that radiation passes through it can be accurately calculated and the 'open-to-shut' time ratio calculated. Such a device was described by Jones and Clarke [27] who pointed out two important considerations. The detector system must average the pulsed flux correctly (i.e. it must obey Talbot's law) and the method of timing the 'open-to-shut' ratio must be capable of very high accuracy and precision. These devices are not commonly used with modern commercial double-beam spectrophotometers, as the chopping rate could interfere with the normal chopping cycle of the spectrophotometer.

5.6 Metallic filters

One of the major drawbacks of conventional glass neutral density filters is, of course, the fact that they cannot be used in the UV, as all known neutral absorbing glasses have a cut-off edge just below the visible range. Evaporated metal-on-quartz or fused silica has therefore been investigated by a number of workers, notably Mavrodineanu [28] at the NBS and Clarke at the NPL [29]. The NBS papers are brought together in a NIST special publication [30]. The major limitation of such filters is the inter-reflections which can occur in instruments, resulting from their intrinsic property of attenuating radiation by reflection. However, this apparent drawback can be put to good use, as will be described later.

In his paper, Mavrodineanu describes results obtained with filters of Inconel-on-fused-silica, with the Inconel protected by a clean fused silica plate held in place with an organic cement. Initial results with the NBS high-accuracy spectrophotometer indicated that a positioning error of 3° about the vertical axis could be tolerated. Subsequent evaluation in commercially available spectrophotometers indicated that the filters could be acceptable as transfer standards in spectrophotometry.

Further work by Clarke and co-workers showed that, by carefully using a particular hard variety of Nichrome alloy, no cover glass is required. Absorbance curves for this material are shown in Fig. 5.7. Also shown is a typical NG-3 filter, and it will be seen that there are a number of points where the transmission profiles intersect. Clarke has shown that these points can be used as diagnostic aids as follows.

Let D_m and D_a be the respective instrument errors in absorbance observed for metal film and absorbing glass filters of nearly equal

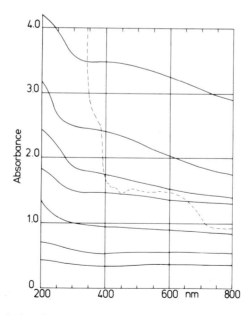

Fig. 5.7 Typical absorbance spectra for NPL Nichrome-on-fused-silica filters (———) compared with a Schott NG-3 filter (- - - -) [29].

absorbance, and let $R_{(m,a)}$ be the ratio of the metal film reflectance to the reflectance of the absorbing glass filter. The inter-reflective component D_{ra} of the observed error with the absorbing glass filter is then given by:

$$D_{ra} = \frac{D_m - D_a}{R_{m,a} - 1}$$

$R_{(m,a)}$ will typically be in the range 5–15 for all Nichrome filters of absorbance greater than 0.4. The basic photometric scale error is then simply $D_a - D_{ra}$. A fuller description of the theory and application is given in Ref. [29]

5.7 Light addition methods

In the previous sections, various means of attenuating the beam were considered. All, however, suffer from some deficiency or other, many, e.g. rely on another laboratory having initially calibrated the absorbing material. An alternative approach that is simple, at least in theory, is the light addition methods.

If there are two radiant fluxes X and Y passing through a spectrometer which individually give readings $I(X)$ and $I(Y)$, then when the two fluxes are added, a reading $I(X+Y)$ will be obtained. If the expression

$$\frac{I(X) + I(Y)}{I(X+Y)}$$

is not equal to unity, errors exist and the system is non-linear.

The technique was probably first used by Elster and Geilet [31] in 1893, and many workers have since produced variations on the theme [32–35]. The most important of these variations are: (i) supplementary light methods as used by Reule [32]; and (ii) aperture methods using various apertures that are opened and closed separately or in combination, as used by Clarke [33], Douglass and Emary [34], and Mielenz and Eckerle [35].

Of these, the second approach is probably preferable in that a second light source is not required and the apparatus can be relatively simple, a schematic diagram of such a device is shown in Fig. 5.8. These devices form the basis of the checking carried out by both the NPL and NIST on their high-accuracy spectrophotometers before calibrating and issuing calibrated Schott filters. A device working on a very similar principle had been developed some years previously by Gilford Instruments [34]. This involved two light-integration areas, coupled by ten apertures, so giving a large range of calibration points. A similar device is also available commercially from Varian for their range of spectrometers. In general, these techniques are probably beyond the capabilities and needs of most

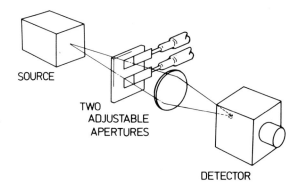

Fig. 5.8 Schematic diagram of the double-aperture method.

users, and the excellent papers produced both by the NPL and NBS should be consulted for further details [29, 30].

5.8 Conclusions and recommendations

For routine use, the recommendation must be a set of neutral Schott glasses, as supplied by NPL or NIST, or by a number of the major instrument manufacturers. These should be submitted for recalibration at yearly intervals.

Alternatively, evaporated metal-on-quartz filters may be used. These have the advantage that they can be used to check absorbance accuracy in the UV as well as the visible region. Their main advantage, however, lies in their use in tracking down the effects of back-reflections within the instrument.

Finally, comment should be offered on the relative merits of solutions versus solid filters. The author must agree with the comments expressed by Clarke [27]: 'The liquid standard tests the complete organization in a chemistry laboratory, if the solution is to be made up to a specified concentration and purity. On the other hand, solid spectrophotometric standards, if stable, can be used to test specifically the intrinsic performance of the instrument, if conditions of use are tightly controlled ...' 'If one wants to consider the actual photomeasurement capability of the instrument apart from considerations of support-staff competence, liquid and cuvette handling procedures or the quality of the cuvettes involved, then solid transmittance standards are the indicated choice'.

References

1. Gibson, K.S. (1949) NBS Circular 484.
2. Gibson, K.S., Walker, G.K. and Brown, M.E. (1934) *J. Opt. Soc. Amer.*, **24**, 58.

3. NBS Circular LCIO17 10 (1955).
4. Gibson, K.S. and Balcom, M.M. (1947) *J. Res. Nat. Bur. Std.*, **38**, 601; also (1947) *J. Opt. Soc. Amer.*, **37**, 593.
5. Copeland, B.E., King, J. and Willis, C. (1968) *Amer. J. Clin. Path.*, **49**, 459.
6. Rand, R.N. (1969) *Clin. Chem.*, **15**, 839.
7. Slavin, W. (1962) *J. Opt. Soc. Amer.*, **52**, 1399.
8. Porro, T.J. and Morse, H.T. (1966) *Pittsburg Conference on Analytical Chemistry*.
9. NIST Special Publication 260–116. *Glass Filters as a Standard Reference Material for Spectrophotometry – Selection, Preparation, Certification and Use of SRM 930 and SRM1930.*
10. Messman, J.D. and Smith, M.V. (1991) *Spectrochimica Acta*, **46B**, 1653–1662.
11. Verrill, J.F. in *Advances in Standards & Methodology in Spectrophotometry*, eds Burgess, C. and Mielenz, K.D., Elsevier, Amsterdam, 1987, 111–124.
12. Verrill, J.F. (1995) In: *Advances in Standards & Methodology in Spectrophotometry* (eds C. Burgess and D.G. Jones)., pp. 49–65. Elsevier Science, Amsterdam.
13. Sharpe, M.R. (1975) *UV Spec. Grp. Bull.*, **3**, 57.
14. Popplewell, B.P. (1975) *Measurement Focus (British Calibration Service Newsletter)*, **6**, 1.
15. Popplewell, B.P. (1977) *UV Spec. Grp. Bull*, **5**, 90.
16. Travis, J., Private communication.
17. Mavrodineanu, R. and Baldwin, J.R. (1975) NBS Special Publication 260–51.
18. Mielenz, K.D. and Mavrodineanu, R. (1973) *J. Res. Nat. Bur. Std.*, **77A**, 699.
19. Dobrowlski, J.A., Marsh, G.E., Charbonneau, D.G., Eng, J. and Josephy, P.D. (1977) *Appl. Opt.*, **11**, 594.
20. Bennett, H.E. (1966) *Appl Opt.*, **5**, 1265.
21. Mielenz, K.D. and Eckerle, K.L. (1972) *Appl. Opt.*, **11**, 594.
22. Heidt, L.J. and Bosley, D.E. (1953) *J. Opt. Soc. Amer.*, **43**, 760.
23. Newman, P.A. and Binder, R. (1961) *Rev. Scl. Instr.*, **32**, 351.
24. Vanderbelt, J.M. (1962) *J. Opt. Soc. Amer.*, **52**, 284.
25. Slavin, W. (1962) *J. Opt. Soc. Amer.*, **52**, 1399.
26. Bryan, F.R. (1963) *Appl. Spectr.*, **17**, 19.
27. Jones, O.C. and Clarke, F.J.J. (1951) *Nature*, **191**, 1290.
28. Mavrodineanu, R. (1976) *J. Res. Nat. Bur. Std.*, **80A**, 637.
29. Clarke, F.J.J. (1977) *UV Spec. Grp. Bull.*, **5**, 104.
30. NBS Special Publication 260–68. *Metal on Quartz Filters as a Standard Reference Material for Spectrophotometry.*
31. Elster, J. and Geilet, H. (1893) *Wied. Ann.*, **48**, 625.
32. Reulc, A. (1968) *Appl. Opt.*, **7**, 1023.

33. Clarke, F.J.J. (1972) *J. Res. Nat. Bur. Std.*, **76A**, 375.
34. Douglass, S. and Emary, R. (1976) *Lab. Equip. Digest*, December, 57.
35. Mielenz, K.D. and Eckerle, K.L. (1972) *Appl. Opt.*, **11**, 2294.

6 Stray-light

6.1 Introduction

Stray-light and poverty, although undesirable, will always be with us because both are relative to the prevailing accepted standard for each phenomena at any given time. What passes for poverty in the developed world today would be regarded as affluence 40 years ago, and likewise a relative stray-light level in UV/VIS dispersive spectrophotometry of 0.1% today is frowned upon but would be specified for only the most advanced dispersing spectrophotometer on the market after World War II [1]. Edisbury [2] aptly paraphrased stray-light for spectroscopists, via the Shakespearean quotation, as 'company, villainous company, hath been the spoil of me' [Henry IV(I) iii 10]. Burgess [3] succinctly defines stray-light as 'radiant energy to which the detector is sensitive, and which should not be there'. Mielenz *et al.* [4] were more pedantic in defining stray-light as 'spurious radiant energy that has departed from its regular path in a spectrophotometer and then re-enters the path so that it is sensed by the detector and causes false readings of transmittance and absorbance'. Henceforth, 'stray-light' will be referred to herein as 'stray radiant energy', abbreviated to SRE. The early editions of the handbooks issued with first generation spectrophotometers made no direct reference to SRE [5]. 'Scattered light' was referred to by a Mr Donaldson during the discussion which took place on Beaven's [6] talk on the Beckman Model DU spectrophotometer given at the inaugural meeting of the Photo-electric Spectrometry Group at Cambridge in July 1948, indicating that users of instrumentation are usually the first to detect shortcomings in the application of an instrument, and users of spectrophotometers are no exception to this rule. Manufacturers became obliged to specify instrument performance in ever increasing detail as time went by. Goldring [7] acknowledged J.R. Edisbury's observation on the effect that 'scattered light' had on underestimating high absorbances. An indication of the awareness of SRE among spectroscopists in 1950 is the fact that the Photoelectric Spectrometry Group Bulletin No. 3 [8–11] was given over entirely to the topic. The specification sheets for spectrophotometers are still parsimonious with information on SRE. The Shimadzu UV-160A specifies the SRE at 340 nm and 220 nm to be < 0.1% using a UV-39 filter and a $10\,\text{g}\,\text{l}^{-1}$ NaI solution, respectively, while the Cary 3E UV-VIS spectrophotometer specifies the SRE at 220 nm to be $< 0.0005\%$ ($10\,\text{g}\,\text{l}^{-1}$ NaI ASTM method) and at 370 nm to be $< 0.0002\%$ ($50\,\text{g}\,\text{l}^{-1}$ NaNO$_2$). A SRE value that is stated without reference to the test procedure is of limited value. Manufacturers may be forgiven for not being more specific on SRE because it is such a will-of-the-wisp quantity, since it may vary from wavelength to wavelength within an instrument, it may vary with time at a given wavelength for an instrument and it may vary from instrument to instrument of the same type, etc. Burgess [3] has adverted to the fact that SRE which lies outside the cutoff wavelength of a detector will not be detected. Buist [12] has stated that SRE increases slowly with time due to deterioration of the source and optics, but that instrumental light leaks-cum-sample fluorescence if present can be easily detected and eliminated as contributing sources to SRE.

SRE causes errors in the determination of absorbance thereby giving rise to increasing deviation from the Beer–Lambert law at ever increasing absorbance values [13–23]. Double-beam instrumentation does not compensate for SRE errors as it does for source emission fluctuation errors, because the effect of SRE is additive as distinct from multiplicative in the sample and reference beams. Irish and Brickell [24] highlight how double monochromation reduces SRE in dispersive UV–VIS spectrophotometers. SRE being no stranger to dispersive optical-null and ratio-recording IR spectrophotometers [25, 26] is in theory absent from Fourier transform infra-red (FT-IR) spectrophotometers. The Perkin-Elmer model 983 IR spectrophotometer has a specified relative SRE level of $< 0.1\%$ at $4000\,\text{cm}^{-1}$ and 0.4% at $400\,\text{cm}^{-1}$. However, Tripp and McFarlane [27] have shown that the high SRE rejection efficiency of FT-IR may not be valid for heated samples, because some sample emission makes its way back to the interferometer, where it is modulated, as was emission from the IR source. Buist [12] suggests that SRE will remain a problem until cheap tunable LASERs become available, and until then discussions on the origins, determination, minimization, etc. of the same must not be forsaken. However, Paul and Saykally [28] write that O'Keefe and Deacon [29] have developed cavity ringdown laser absorption spectroscopy (CRLAS) using tunable pulsed laser sources.

6.2 Definitions

SRE, being of wavelengths outside the narrow band nominally transmitted by the monochromator, contributes to the overall photosignal in a complex way owing to the detector's sensitivity being wavelength dependent. Optical designers may be interested in the spectral composition of the SRE emanating as monochromator stray-light (MSL) with the intention of tracing its origins so that the optical components and design of monochromators might be improved to minimize the SRE, but analytical scientists have to live with the total instrument and are more concerned with the lumped effect of all SRE emanating as instrumental stray-light (ISL). Slavin [30] defined relative SRE ('s') to be 'the ratio of the signal ('S') produced by the detector for radiation of all wavelengths outside the monochromator spectral slitwidth, to the total signal ('I_0') at a particular wavelength setting', i.e., $s = S/I_0$. The absolute amount of SRE present is less relevant than the ratio defined by Slavin in determining the effect on absorbance readings.

6.3 Origin of SRE

An ideal monochromator transmits radiation only within its spectral slitwidth, while all other wavelengths entering the monochromator are absorbed. However, scattering and diffraction from imperfections in mechanically blazed grating surfaces inside the monochromator generate MSL and contaminate the monochromatic exit beam. Sharpe and Irish [31], and Irish and Brickell [24] have indicated that groove depth variation, groove spacing errors and groove straightness give rise to MSL, and that blazed holographic gratings produced by photoresist methods from laser-generated interference patterns give an order of magnitude less MSL than mechanically blazed gratings. Nearly all spectrophotometers use replica gratings, copies of master gratings (ruled or holographic), as the spectral dispersing element made by a liquid resin casting process which preserves the optical accuracy of the master gratings to perfection [22]. There is no optical test that can distinguish between master and replica no matter whether the master was mechanically ruled or made in photoresist by holographic techniques [32, 33]. The MSL and monochromatic radiant intensity have little dependences on slitwidth and slit height, therefore, the contributions of MSL to the relative SRE level intensity have like dependences on slitwidth and slit height, and therefore, the contribution of MSL to the relative SRE level will be independent of slit dimensions [12, 21]. When working in the spectral region 320–400 nm common to both the deuterium and tungsten lamps, the D_2-lamp contributes relatively less MSL than the W-lamp, unless emission from the latter at wavelengths greater than 400 nm is filtered out.

The MSL and monochromatic radiation emanating from the monochromator may be further modified by other components in the spectrophotometer, and Kaye [34] has written that the presence of any test material in a spectrophotometer will always reduce the relative SRE level by absorbing it, as does the aging of optical surfaces particularly in the response to UV radiation. SRE becomes a serious problem at the wavelength limits of a UV–VIS spectrophotometer where signal energy is at a minimum [21, 35]. This is the reason why manufacturers quote SRE at

220 nm where the D_2-lamp energy is small, and also at 340 nm close to a minimum of the W-lamp energy, and 340 nm is also a widely used biochemical wavelength [21]. At the short wavelength limit (< 300 nm), although detector sensitivity may be excellent, the D_2-lamp emission and optical window-cum-solvent transmittance falls away, thereby allowing the ISL of longer wavelengths to swamp the monochromatic signal. At the long wavelength limit (> 700 nm), reduced detector sensitivity allows shorter wavelength ISL to swamp the monochromatic signal even though the monochromatic radiant intensities from W-lamps and optical window-cum-solvent transmittance are more than adequate. The ISL can be significantly reduced in the above cases by inserting stray-light filters with appropriate 'cut-off' characteristics.

6.4 Stray radiant energy errors

The definition, origin and effects of SRE on spectrophotometric measurements have been reviewed by Zhu and Qian [36, 37]. The relative SRE level is a function of the sample, and so a general statement of the transmittance error introduced cannot be made, although the SRE in the absence of the sample is known. The relationship between the monochromatic transmittance (T_λ) of a sample, the observed transmittance (T') and the relative SRE level (s), which was derived from first principles [38], cf. below, will emphasize the point. It is assumed that the cuvette wall and blank materials are transparent to all radiations in question.

I_0 = incident monochromatic light intensity,
I_λ = transmitted monochromatic light intensity,
$T_\lambda = I_\lambda/I_0$ = monochromatic transmittance of sample,
S = incident SRE intensity,
$s = S/I_0$ = relative SRE level $<<1$,
St^A = transmitted SRE intensity,
$A = -\log_{10} T_\lambda$ = monochromatic absorbance of the sample,
t^A = transmittance of sample to SRE,

$$T' = \frac{I_\lambda + St^A}{I_0 + S} = \frac{T_\lambda + st^A}{1+s}$$
$$= (T_\lambda + st^A)(1+s)^{-1} = (1-s)T_\lambda + t^A(s-s^2)$$
$$= T_\lambda + s(t^A - T_\lambda) \tag{6.1}$$

An inspection of Equation (6.1) indicates that the observed transmittance may be greater, equal to or less than the monochromatic transmittance, depending on whether 't', the sample's transmittance to SRE, is greater, equal to or less than 0.1. The most likely situation involves $t > 0.1$, i.e. $t^A > T_\lambda$, meaning that the observed absorbance (A') will be less than the monochromatic absorbance (A), cf. Equation (6.2). The situation $t < 0.1$, giving $A' > A$, is unusual but may be encountered when making measurements near absorbance minima. However, it is often assumed that $t = 1$.

$$A' = -\log_{10} T' = -\log_{10}(10^{-A} + s(t^A - 10^{-A})) \quad (6.2)$$

Table 6.1 shows a range of calculated values of A' for $t = 0.8$, and $s = 10^{-2}$, 10^{-3}, 10^{-4} and 10^{-5}. Figure 6.1 shows plots of A' versus A for the data in Table 6.1. SRE can distort a recorded absorbance peak if the relative SRE level is changing in the spectral region of the peak [12]. This may arise through a progressive change in the monochromatic signal level with wavelength while the signal level due to ISL remains constant. The said progressive change in the monochromatic signal level with wavelength may be due to changing solvent absorption over the same wavelength range. Figures 6.2 and 6.3 show the absorbance spectra of maleic acid of various concentrations in ethanol and water, respectively [12]. Contrast the shifting peak maximum, peak asymmetry and increased noise level in Fig. 6.2 with the 'normal' peak characteristics in Fig. 6.3. Figure 6.4 shows that the absorbance spectrum of ethanol against water 'blank' becomes strongly absorbing below 215 nm. The narrower the absorbance peak the more it will be affected by SRE.

6.5 SRE reduction

The relative SRE level may be decreased either by increasing the energy throughput of the instrument at the wanted wavelength relative to the throughput at the unwanted wavelengths, or by reducing the scattering of

Table 6.1 The effect of SRE upon the observed absorbance (A') of a sample, assuming that the transmittance of the sample to SRE is 0.8.

Monochromatic absorbance (A_λ)	Relative SRE level			
	10^{-2}	10^{-3}	10^{-4}	10^{-5}
	Observed absorbance (A')			
0.300	0.296	0.300	0.300	0.300
1.000	0.971	0.997	1.000	1.000
1.300	1.244	1.294	1.300	1.300
2.000	1.788	1.973	1.997	2.000
2.300	1.961	2.252	2.295	2.300
3.000	2.214	2.821	2.978	2.998
3.300	2.277	3.009	3.261	3.296
4.000	2.377	3.293	3.851	3.983

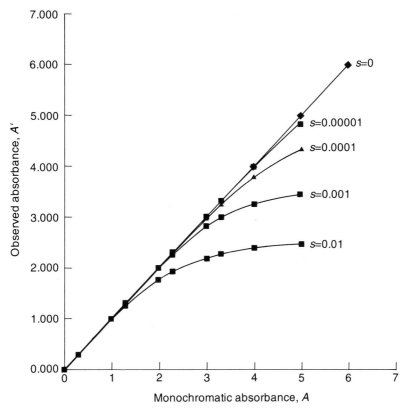

Fig. 6.1 Observed absorbance plotted against monochromatic absorbance according to Equation (6.2) for $t = 0.8$ and $s = 0$, 10^{-5}, 10^{-4}, 10^{-3} and 10^{-2}.

unwanted wavelengths in the monochromator. In the spectral range between 320 and 390 nm it is possible to use a W-lamp or D_2-lamp. Relative SRE will be lower for the latter lamp in the absence of any other SRE amelioration technique being used. However, most spectrophotometers employ a SRE filter, e.g. Corning 9863, Schott UG5 or Hoya U-330 [12], in the light beam below 400 nm with a W-lamp, because it absorbs most of its visible emission but is transparent to its near-UV emission. If the spectrophotometer is set to the very important biochemical wavelength of 340 nm, then most of the SRE from the W-lamp, being visible, is absorbed by the filter. A filter material, transparent below 220 nm, if discovered would be very useful.

The upper wavelength limit of many modern spectrophotometers has been extended to 1100 nm with the deployment of detectors which are sensitive in this spectral region, e.g. the Shimadzu 160. However, in a grating-based instrument when the wavelength is set to transmit at λ, there will be some radiation of wavelengths, $\lambda/2$ in the second order and $\lambda/3$ in

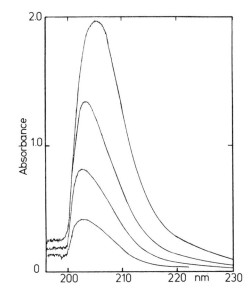

Fig. 6.2 [12] The absorbance spectra of maleic acid in ethanol 'blank' at various concentrations, showing the effect of the increasing relative SRE level with decreasing wavelength due to the ethanol 'blank' becoming strongly absorbing below 215 nm.

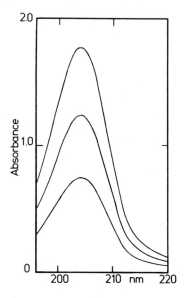

Fig. 6.3 [12] The absorbance spectra of maleic acid in water 'blank' at various concentrations.

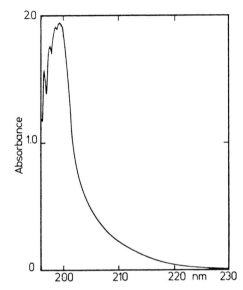

Fig. 6.4 [12] The absorbance spectrum of ethanol in water 'blank' showing the effect of SRE.

the third order, etc. transmitted as well. These unwanted wavelengths may be excluded by inserting an order-sorting glass filter automatically at a wavelength of approximately 700 nm. This red-transmitting (blue-absorbing) filter stops second and third order wavelengths between 350 and 550 nm being detected. 'Order sorting' is not a problem at wavelengths below 700 nm because the shorter wavelength output from W-lamps is very low, and neither is it a problem at wavelengths below 380 nm when a D_2-lamp is employed because the optical materials involved tend to be opaque to wavelengths below 190 nm. The number of such filters employed (order sorting and SRE types) and the wavelengths at which their insertion takes place differs from instrument to instrument, and is dependent on monochromator design and the blazed characteristics of its diffraction grating. The Shimadzu 260 betrays the change over wavelengths by momentarily pausing during a scan to perform the switch. The Perkin-Elmer 551S employs a continuously rotating filter wheel as it scans, and the change over is smoother. The step-like filter change over manifests itself as a sudden reduction in the relative SRE level after the switch has occurred, cf. Fig. 6.5, which shows the differential absorbance spectrum of 10 mm versus 5 mm concentrated red food-dye solution scanned in a Shimadzu 260 spectrophotometer at 5 nm spectral slitwidth. The SRE 'peaks' of absorbance 1.56 and 1.68, at 565 nm and 455 nm, respectively, indicate relative SRE levels of 2×10^{-4} and 1×10^{-4}, respectively, for the cell path thickness ratio of 2 [4, 39, 40]. Note the 'step' increase of 0.12 in the recorded differential absorbance at 531.4 nm where

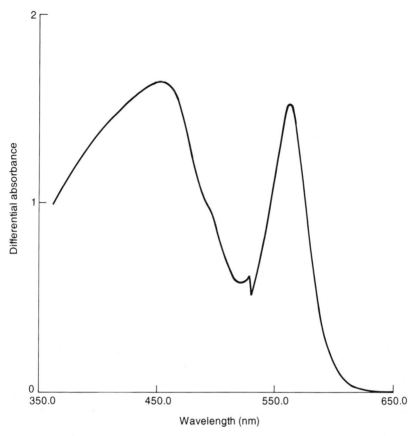

Fig. 6.5 Mielenz-type differential absorbance spectrum of 10 mm versus 5 mm concentrated red food-dye solution scanned in a Shimadzu 260 spectrophotometer at 5 nm slitwidth.

a pause for a filter change occurred during the scan, corresponding exactly to the difference between the differential absorbance maxima noted above. The filter change over halved the relative SRE level.

Figure 6.6 shows separately the spectral energy profiles of the individual D_2- and W-lamps in a Shimadzu 260 spectrophotometer. The detection sensitivities may have differed from one spectral profile to the other and, therefore, are not directly comparable.

6.6 SRE measurement

The measurement of MSL is a tedious process [12] requiring specialized equipment, and only a few such measurements have been reported [9, 41, 42]. However, it is normally sufficient to measure the overall effect of the

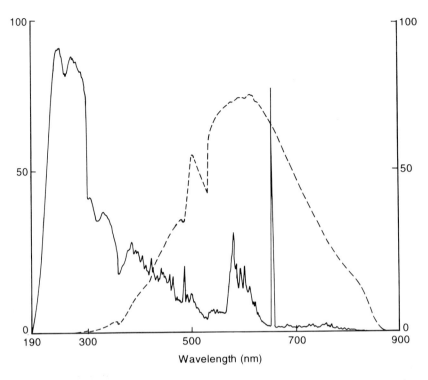

Fig. 6.6 The emission profiles of the D_2 (———) and W (- - - -) radiant sources in a Shimadzu 260 spectrophotometer. This represents a combination of lamp outputs, mirror reflectivities, grating efficiency, SRE order-sorting filter transmittances and detector spectral response.

SRE in the instrument manifested as deviations in its absorbance readings. The apparent absorbance of an edge-absorbing filter or material just beyond its cut-off wavelength was recommended by ASTM [20] and Burgess [3] among others [12, 16, 35] to determine SRE. Rewriting Equation (6.1) for $T_\lambda \ll T'$ gives $st^A = T' = 10^{-A'}$, where A' is the apparent absorbance measured at wavelength λ in Fig. 6.7. The relative SRE level is checked at a specific wavelength by selecting a cut-off solution for that wavelength. Place the solution in the correct pathlength cell and scan down from a wavelength above the edge absorbance wavelength. The absorbance should increase step-wise to A_{fsd} at the cut-off wavelength and maintain that reading at all shorter wavelengths. However, if SRE is present, the reading will not attain A_{fsd} but will form an asymmetrical peak. This method is applicable on the assumptions that $t = 1$ and $A'_{max} < A_{fsd}$ of the instrument. If the latter assumption is not fulfilled, then it may be remedied by setting the 100% T with a sufficiently absorbing 'blank' (A_b) in order to keep the apparent absorbance reading (A'') on scale, and then $A' = A'' + A_b$. An example of this is shown by

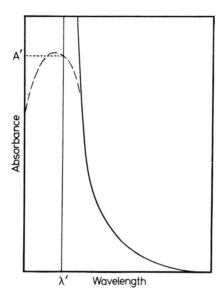

Fig. 6.7 [12] The effect of SRE on the observed absorbance, A', (- - - -) of a cut-off filter.

Mielenz et al [4] in Fig. 6.8, where the cut-off filter used is a $10\,\text{g}\,\text{l}^{-1}$ KI solution in a 10 mm cell. The instrument under test was a double beam recording spectrophotometer with $A_{\text{fsd}} = 3$. Two attenuating filters of blank absorbances 1.92 and 1.13, respectively, were required to observe the plateau absorbance of 2.37. Hence, the apparent absorbance, $A' = 5.42$, and the relative SRE level was 3.8×10^{-6}. Since $t < 1$, the measured relative SRE level is an underestimate. It was necessary in this instance to 'back-off' the reference beam's absorbance by at least 2.42 so as to keep the sample beam's absorbance on scale. Edge-absorbing filters and materials are listed by several authors, e.g. Irish and Brickell [24], ASTM/E-387-84 [20], Mielenz et al. [4], Slavin [30], Francis [22], and Hartree [16], cf. Table 6.2. Figure 6.9(a) shows the transmittance spectrum of a $10\,\text{g}\,\text{l}^{-1}$ NaI aqueous solution, and Fig. 6.9(b) shows the true (———) and observed (- - - -) absorption spectra of the same solution recorded with a grating-based spectrophotometer.

A novel variation of the same method was proposed by Mielenz et al.[4], who simplified it by attenuating the reference beam with the same cut-off filter solution placed in a 5 mm cell and using a 10 mm cell in the sample beam. A maximum differential absorbance $\Delta A'$ is observed, while scanning through the edge absorbance, which is related to the relative SRE level by $s = 0.25 \times 10^{-\Delta A}$. The resultant trace, shown in Fig. 6.10, has $\Delta A = 2.36$ which suggests a relative SRE level of 4.8×10^{-6}, which compares favourably with the value given above for the same instrument but

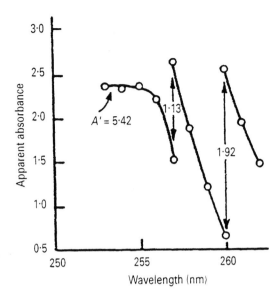

Fig. 6.8 [4] Stray radiant energy ASTM test method E387 on a single-beam double-monochromator spectrophotometer, with a 0–3 absorbance scale, using KI cut-off filters near 255 nm. The ASTM method required successive 'blanking' with two neutral density screens with attenuations 1.92 and 1.13 absorbance units, respectively. (Reproduced with permission of Academic Press.)

Table 6.2 Cut-off filters for SRE tests in the VUV/UV/VIS/NIR/IR spectral regions courtesy of ASTM [20], and Irish and Brickell [24].

Spectral range (nm)	Filter material	Pathlength (mm)
165–173.5	Water	0.1
170–183.5	Water	10.0
175–200	KCl(12 g l^{-1} aq.)	10.0
195–223	NaBr(10 g l^{-1} aq.)	10.0
210–259	NaI(10 g l^{-1} aq.)	10.0
250–320	Acetone	10.0
300–385	NaNO$_2$(50 g l^{-1} aq.)	10.0
600–660	Methylene Blue (0.005% aq.)	10.0
1660–1750	CH$_2$Br$_2$	50.0
2050–1200 cm^{-1}	Fused silica	2.0
1140–800 cm^{-1}	LiF	6.0
760–600 cm^{-1}	CaF$_2$	6.0
630–400 cm^{-1}	NaF	6.0
410–250 cm^{-1}	NaCl	6.0
240–200 cm^{-1}	KBr	6.0

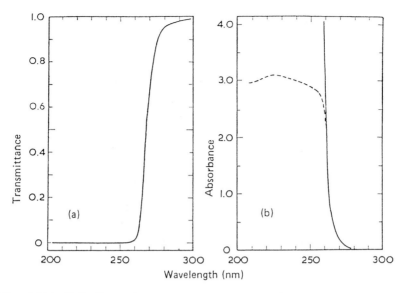

Fig. 6.9(a) [24] Transmittance spectrum of $10\,g\,l^{-1}$ NaI aqueous solution measured in a 10 mm cell. (b) [24] True absorbance (———) and observed absorbance (- - - -) spectra of $10\,g\,l^{-1}$ NaI aqueous solution measured in a 10 mm cell by a typical laboratory spectrophotometer. The observed absorbance of 3.08 at 225 nm means that the relative SRE level at that wavelength is 0.00083 and consists of wavelengths greater than 265 nm.

for a different method. Comparative data for 15 instruments showed good agreement between the said methods, cf. Fig. 6.11. The slope of the line in Fig. 6.11 was 1.01 ±0.01 and Mielenz et al. [4] conclude that both methods give equivalent results.

Fleming [40] generalized the Mielenz et al. variation [4] by deriving an exact equation relating the relative SRE level present to the observed differential absorbance maximum, and for all possible cell pathlength ratios, and made it applicable to double-beam ratio-recording [26, 39, 40], as well as double-beam optical-null spectrophotometers [25, 43]. Transmittance ratio spectrophotometry involves preparing a geometrical concentration series of solutions, and the transmittance of a solution relative to its less concentrated nearest-neighbour is called its transmittance ratio. Experimental observations have shown [44, 45] that the transmittance ratio if plotted against the monochromatic absorbance of the reference solution will initially decrease from unity, as expected from Beer–Lambert considerations, but as the reference absorbance increases and contrary to theoretical expectations, the transmittance ratio will 'bottom out' and commence to increase again. Referring to Fig. 6.12, the theoretical curve for zero SRE is curve A, while curve B shows the experimentally determined curve. The plot was determined as follows:

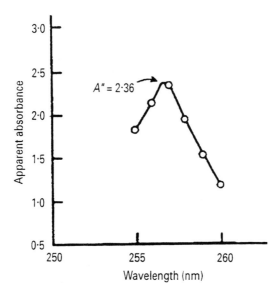

Fig. 6.10 [4] The Mielenz SRE test method on a single-beam double-monochromator spectrophotometer with a 0–3 absorbance scale is shown. The spectrophotometer was 'blanked' in turn at each wavelength with 5 mm of $10\,\mathrm{g\,l^{-1}}$ aqueous KI, and the differential absorbance at each successive wavelength of 10 mm of $10\,\mathrm{g\,l^{-1}}$ aqueous KI was measured. (Reproduced with permission of Academic Press.)

(a) From a stock solution having a monochromatic absorbance between 4 and 5, dilutions were prepared at a constant geometric ratio of 1.25 by diluting 20 ml to 25 ml. The geometric concentration series will also form a geometric monochromatic absorbance series.

(b) Blank the spectrophotometer with the least absorbing solution and measure the transmittance ratio of the next most absorbing solution, etc.

(c) Plot the transmittance ratio versus the monochromatic reference absorbance. It has been shown [39] that the relative SRE level s is given, for $\alpha = 1.25$ and $\rho = 0.345$, by:

$$s = [1 + (\alpha - 1)^{-1}(1 - \rho)(\rho/\alpha)^{\alpha/(\alpha - 1)}]^{-1}$$
$$= [1 + 4(1 - \rho)(0.8\rho)^{-5}]^{-1} = 6.1 \times 10^{-4} \tag{6.3}$$

Figure 6.13 [3] shows the effect of various values of relative SRE level on the transmittance ratio minimum as calculated from Equation (6.3) for $\alpha = 1.25$. Figure 6.14 [46] graphically shows the dependence of the transmittance ratio minimum on the corresponding reference beam's monochromatic absorbance ($= -\log_{10}\tau$) in the relative SRE range $0.1 > s > 0.00005$ and for $1.25 < \alpha < 10$. This may be utilized to readily calculate the

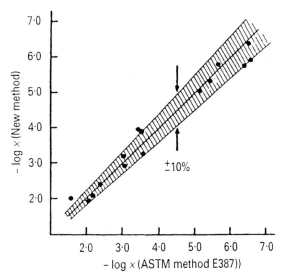

Fig. 6.11 [4] Comparative results obtained in 15 SRE tests by the ASTM test method E387 [20] and the Mielenz et al. [4] test method showing the equivalence of both methods within 10% limits. (Reproduced with permission of Academic Press.)

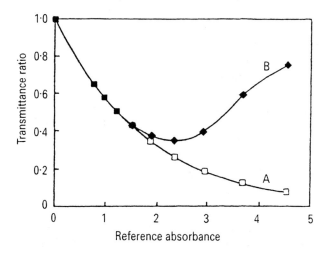

Fig. 6.12 [44] Theoretical curve (A) and experimental curve (B) showing the dependence of the transmittance ratio on the monochromatic reference absorbance for a concentration ratio of $\alpha = 1.25$. (Reproduced with permission of Academic Press.)

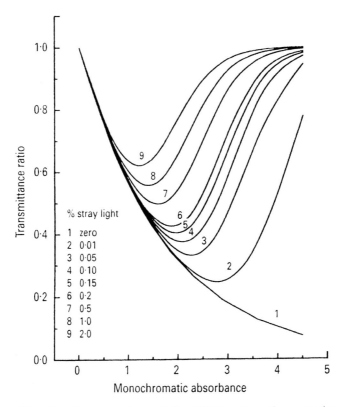

Fig. 6.13 [3] The effect of various relative SRE levels on the transmittance ratio minimum, ρ, for $\alpha = 1.25$ inserted into Equation (6.3). (Reproduced with permission of Academic Press.)

relative SRE level once the transmittance ratio minimum has been determined for a given α-value. The transmittance ratio SRE test methods of Fleming [39] and Mielenz et al. [4] (also reported by Mavrodineanu and Burke [47]) are related. The latter method gradually increases differential absorbance by wavelength scanning through the leading or trailing absorbance edge of a cut-off filter/solution which has been placed in both the sample and reference beams in a definite thickness/concentration ratio, whereas the transmittance ratio method is carried out at a fixed wavelength and the differential absorbance is increased by gradually advancing the concentration of the solutions held in the sample and reference beams. The differential absorbance for both methods will not increase indefinitely, but will peak at a value determined by both the SRE level present in the instrument and the concentration/thickness ratio in use. The two test methods may be regarded as opposite sides of the same coin. Although the equations dealing with the transmittance ratio sample-based method have been formulated assuming that the SRE is absorbed

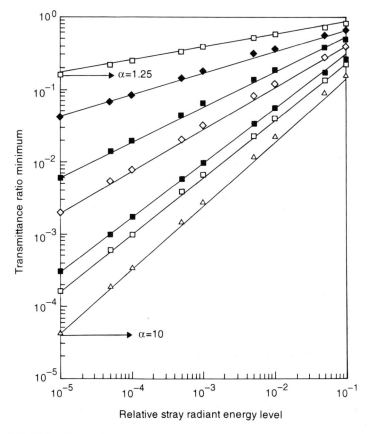

Fig. 6.14 [46] Transmittance ratio minimum, ρ, versus relative SRE level, s, plotted on log–log axes for $\alpha = 1.25$, 1.5, 2, 2.5, 4, 5 and 10.

by the sample [25, 26, 40, 43], there is no direct way of determining the same from the measurements taken. Both approaches are likely to underestimate the relative SRE level present because of sample absorption of the same.

The transmittance ratio test methods involve photometric measurements which are subject to error (cf. Gridgemann [48]). Inspection of equation (6.3) suggests that there is a connection between the photometric error and the error in determining the relative SRE level. Fleming and Fleming [46] investigated the photometric error contributions which the relative errors in determining both the cell pathlength ratio, $\Delta\alpha/\alpha$, and the transmittance ratio minimum, $\Delta\rho/\rho$, contribute towards the relative error in determining the relative SRE levels, $\Delta s/s$, in a double beam ratio-recording spectrophotometer. The analysis enabled optimum experimental conditions to be preselected when using the transmittance ratio SRE test method. Figure 6.15 [46] shows the relative error function, $(\Delta s/s)_{rms}$, plotted against the cell pathlength ratio, α, for seven distinct

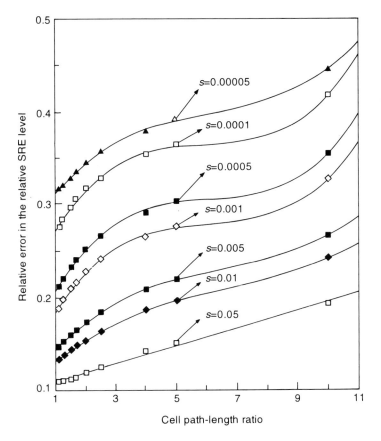

Fig. 6.15 [46] Relative error in the relative SRE level, $\Delta s/s$, plotted against the cell pathlength ratio, α, for several relative SRE levels, $s = 0.05, 0.01, 0.005, 0.001, 0.0005, 0.0001$ and 0.00005.

relative SRE levels. The relative error function increases with increasing α for a given relative SRE level, and increases with decreasing relative SRE levels for a given α. Sample cells come in standard mm-sizes of 1, 2, 5, 10, 20, 50 and 100. Considering the dependence of the error function on α, it is recommended that $\alpha = 2$ be employed, as it may be selected in four independent ways from the above cell size set and it also makes Equation (6.3) convenient to calculate, since:

$$\rho = \alpha t^{\alpha-1} = 2\tau \text{ and } s = (\rho/(\rho - 2))^2 = (\tau/(1 - \tau))^2$$

An *a priori* knowledge of the approximate relative SRE level in a spectrophotometer, e.g. gleaned from its specification sheet, will suggest the monochromatic absorbances of a narrow range of reference beam solutions which need to be prepared for a given α so as to expedite an experimental determination of ρ. Figure 6.14 facilitates the calculation of the reference beam's monochromatic absorbance from the transmittance ratio minimum appropriate to an expected relative SRE level and the α value being used. Spectrophotometers are usually used in the absorbance mode rather than the transmittance mode and therefore, differential absorbance rather than transmittance ratio is observed in this application. Differential absorbance is a slowly varying function of the monochromatic reference absorbance at or near the differential absorbance maximum, i.e. it is 'flat-topped'. The largest α-value permissible is determined by the spectrophotometer's fsd-absorbance and the relative SRE level to be determined, e.g. if $s = 10^{-6}$, then $\alpha = 2.5$ gives $\Delta A_{max} = 3.3$, which may be approaching A_{fsd}. A value of $\alpha = 2$ in similar circumstances gives $\Delta A_{max} = 2.7$.

A direct transmittance spectrometric sample-based method has been developed by Fleming and O'Dea [38] which allows for sample absorption of the SRE. Inspection of Equation (6.2) for $s <<< 1$ [$T' = (1-s)T_\lambda + st^A$] and for the two extreme cases where the monochromatic transmittance term is far greater [$(1-s)T_\lambda >> st^A$] or far less than the SRE term [$(1-s)T_\lambda << st^A$] allows the following approximations to be written, respectively, $T' = T_\lambda$ and $T' = st^A$. Since absorbance ($-\log_{10} T_\lambda$) is the parameter usually observed, the following may be written:

$$\log_{10} T' = \log_{10} T_\lambda = -A_\lambda \text{ for } s < 0.02 T_\lambda \tag{6.4}$$

$$\log_{10} T' = \log_{10} st^A = \log_{10} s + A_\lambda \log_{10} t \text{ for } T_\lambda < 0.02s \tag{6.5}$$

If the observed transmittance, T', is plotted on the $\log_{10}$ ordinate axis against the monochromatic absorbance, A_λ, on the linear abscissa axis, then straight line plots, as in Fig. 6.16, result. An 'elbow' region is observed where $s = T_\lambda$. The true relative SRE level is readily calculated by extrapolation of Equation (6.5) to $A_\lambda = 0$. The antilog of the slope from Equation (6.5) gives the SRE transmittance of the sample whose monochromatic absorbance is unity. In practice, the direct transmittance method may require some reference beam attenuation as in the ASTM method to allow on-scale readings to be observed. The monochromatic absorbances in any sized cell are obtained by simple multiplication from those determined in 1 mm cuvettes. The absorbance values in the 1 mm cuvettes must be such that the SRE has negligible influence on the same. The multiplication factor(s) may be spectrophotometrically determined, or obtained by relying on the manufacturer's stated thickness for the cuvettes employed.

The 'total absorber' method [20] requires a sample with an absorption bandwidth at the wavelength of interest which matches the spectral bandwidth of the monochromator while being transparent at all other wavelengths. Its apparent minimum transmittance will be a good approximation to the relative SRE level in the absence of a sample. However, the UV/VIS regions lack such ideal absorbing materials. The

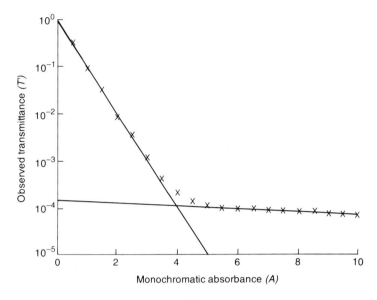

Fig. 6.16 [40] Plot on semi-log axes of the observed transmittance, T', versus the monochromatic absorbance, A_λ, at 654 nm and with a spectral slitwidth of 1 nm. The measurements were made in a pair of matched 20 mm quartz cells with a Shimadzu 260 spectrophotometer. The analyte samples were dilutions of a 50 g l^{-1} Orleans Blue food-dye in distilled water. The monochromatic absorbance range, $7 < A_\lambda < 10$ yielded $s = 0.000140$ and $t = 0.935$.

intense band in the benzene vapour spectrum near 260 nm, hot mercury vapour at 254 nm, and polystyrene films at 13.3 and 14.4 mm are useful if the bands are resolved so as to give reproducible results.

6.7 Method comparison

The relative SRE level at 654 nm and 1 nm spectral slitwidth in a Shimadzu 260 double-beam spectrophotometer was determined by Fleming [40] using three test methods, i.e. that of Mielenz et al [4], transmittance ratio spectrometry [39] and direct transmittance spectrometry [38]. Matched pairs of 1, 2, 5, 10, 20, 50 and 100 mm quartz cells were used, thereby allowing various nominal sample to reference cell pathlength ratios, e.g. 2 (10/5, 20/10), 2.5 (50/20), 4 (20/5) and 5 (50/10). The working solutions were dilutions of a 50 g l^{-1} Orleans blue food-dye (E123) stock acqueous solution. An arithmetic concentration series of the parent was prepared having monochromatic absorbances mm^{-1} at 654 nm in the range $0.025 < A_\lambda$ (mm^{-1}) < 0.5. The arithmetic series had an absorbance increment of 0.025 and this yielded a set of 20 samples (excluding blank) whose monochromatic absorbances ranged from A_{min}

to A_{max} (incremented by steps of ΔA) in various cell pathlengths (b) as follows:

$A_{min} + (\Delta A \times 19 \text{ steps}) = A_{max}$ in b mm pathlength cell
$0.025 + (0.025 \times 19) = 0.5$ in a 1 mm pathlength cell
$0.050 + (0.050 \times 19) = 1.0$ in a 2 mm pathlength cell
$0.125 + (0.125 \times 19) = 2.5$ in a 5 mm pathlength cell
$0.250 + (0.250 \times 19) = 5.0$ in a 10 mm pathlength cell
$0.500 + (0.500 \times 19) = 10.0$ in a 20 mm pathlength cell
$1.250 + (1.250 \times 19) = 25.0$ in a 50 mm pathlength cell

The experimental cell pathlength ratios were determined by measuring the absorbance at a peak wavelength of 630 nm of a dilute Orleans Blue food-dye (E123) solution in all the available cells, and this gave the following relative pathlengths: 1.00 ± 0.008, 2.00 ± 0.013, 5.00 ± 0.017, 10.00 ± 0.013, 20.01 ± 0.013 and 50.00 ± 0.04.

The Mielenz test method was repeatedly applied to the Shimadzu 260 by slowly scanning in the range $750 > \lambda(nm) > 625$, the food-dye test solution having been placed in a pair of matched quartz cells of nominal pathlength ratio $\alpha = 2$ (10 mm versus 5 mm). The ensuing differential absorbance spectrum displayed the expected SRE Mielenz peak at 654 nm. The measured monochromatic absorbance ($= -\log_{10}\tau$) of the test solution at 654 nm in the 5 mm reference cell was 2.00 and the Mielenz peak had an average differential absorbance $\Delta A = -\log_{10}\rho = 1.680 \pm 0.005$. The calculated value for the corresponding monochromatic absorbance $[= -\log_{10}(\rho/\alpha)^{1/(\alpha-1)}]$, is 1.98, which differs from the measured value because allowance has not been made for the absorption of SRE by the test sample. If a more concentrated member of the arithmetic concentrated series had been used in the above Mielenz test, then the SRE peak would have occurred at a longer wavelength, etc. The Mielenz analysis, $s = 0.25 \times 10^{-2\Delta A}$, which is only applicable for $\alpha = 2$, yields $s = 1.09 \times 10^{-4}$ for the above experiment, while Equation (6.3) yields 1.11×10^{-4}. If the Mielenz peak occurs at 654 nm for all α-values, then *a priori* knowledge of the absorbance of the test solution in the reference cell at 654 nm is necessary for each α-value, e.g. if $s = 1.11 \times 10^{-4}$ and $\alpha = 2.5$, then $\rho = 0.0084$ satisfies Equation (6.3). Since $\rho = \alpha\tau^{\alpha-1}$, Orleans Blue food-dye of monochromatic absorbance 1.650 ($= -\log_{10}\tau$) if placed in the reference beam for $\alpha = 2.5$ will give a Mielenz peak at 654 nm, etc. This calculation procedure was repeated for $\alpha = 2.5, 4$ and 5, and was facilitated by having prior knowledge of the absorbance of the Mielenz peak at 654 nm which ensued from scanning the differential absorbance of a selected food-dye sample for $\alpha = 2$. The resulting differential absorbance spectra are given in Fig. 6.17(a) for $\alpha = 2, 2.5, 4$ and 5. The Mielenz peaks

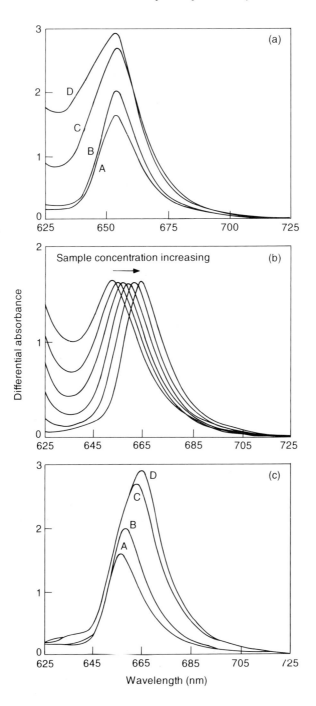

occur at approximately 654 nm for all the α-values used but increase in amplitude with α.

Figure 6.17(b) displays six Mielenz differential absorbance spectra which were obtained for a constant cell pathlength ratio of 2 (= 10/5) and by changing the sample concentration in the cuvettes for each scan. Note the red-shift of the Mielenz peaks which occurs with increasing sample concentration, but the peak/amplitude remains unchanged indicating that the relative SRE level remained constant over a wavelength range of 20 nm.

Figure 6.17(c) displays four Mielenz differential absorbance spectra which were scanned for constant sample concentration in the cuvettes and by changing the cell pathlength ratio between the following nominal values: 2 (= 10/5), 2.5 (= 50/20), 4 (= 20/5) and 5 (= 50/10). Note the red-shift of the Mielenz peaks which occurs with increasing α-values.

Fleming's [39] transmittance ratio, and Fleming and O'Dea's [38] direct transmittance SRE test methods were also applied to the Shimadzu 260 spectrophotometer set at 654 nm and a spectral slitwidth of 1 nm. The ensuing experimental determinations are plotted in Figs 6.18 and 6.16, respectively. The aforementioned food-dye arithmetic concentration series was appropriately employed in both tests. Figure 6.18 displays differential absorbance versus monochromatic reference absorbance plots for four cell pathlength ratios, $a = b_s/b_r$. The experimental cell pathlength ratios are as follows: $20.01/10.00 = 2.001 \pm 0.005$; $50.00/20.01 = 2.499 \pm 0.004$; $20.01/5.00 = 4.00 \pm 0.016$; and $50.00/10.00 = 5.00 \pm 0.01$. Figure 6.15 shows a plot of semi-log axes of the observed average transmittance (derived from the observed average absorbance in 20 mm pathlength cells) of solutions, $T' = 10^{-A'}$, at 654 nm versus the respective monochromatic absorbance (A) in the range $0 < A < 10$. The exponential regression equation to fit to the linear part of the plot in the upper absorbance range is given by: $T' = 0.000140 \times 10^{-0.029A}$, yielding $t = 0.935$ and $s = 0.00014$. The transmittance ratio function (r) which takes into account the absorption of SRE by sample can be numerically modelled via Equations (6.6) and (6.7) below:

Fig. 6.17(a) [40] Four Mielenz et al. differential absorbance spectra in the wavelength range $725 > \lambda(nm) > 625$ for nominal cell pathlength ratios (α_b) of: A, 2; B, 2.5; C, 4; and D, 5. The Orleans Blue food-dye concentration is gradually increased with decreasing α-values so as to maintain the Mielenz peak at 654 nm. (b) [40] Six Mielenz et al. differential absorbance spectra in the wavelength range $725 > \lambda(nm) > 625$ for a nominal cell pathlength ratio (α_b) of 2. (c) [40] Four Mielenz et al. differential absorbance spectra in the wavelength range $725 > \lambda(nm) > 625$ for nominal cell pathlength ratios (α_b) of: A, 2; B, 2.5; C, 4 and D, 5; and for a fixed concentration of Orleans Blue food-dye.

$$r = \frac{T_\lambda^\alpha + st^{\alpha A}}{T_\lambda + st^A} \tag{6.6}$$

$$= \frac{T_\lambda^\alpha + s\mu^\alpha}{T_\lambda + s\mu} \tag{6.7}$$

where μ $(= t^A)$ is the transmittance of the reference sample to SRE. Equation (6.3), which assumes $\mu = 1$, must be modified as in Equation (6.8) below [40]:

$$s = [1 + (\alpha - 1)^{-1}(\mu^\alpha - \mu\rho)(\rho/\alpha)^{\alpha/(\alpha-1)}]^{-1} \tag{6.8}$$

Equation (6.8) cannot be applied to the differential absorbance maxima in Fig. 6.17(a) and (b) without *a priori* knowledge of μ. However, this information may be construed from knowledge of the monochromatic reference absorbance at 654 nm and the value of t defined in Equation (6.1), and calculated as 0.935 from the slope of the linear portion of the upper absorbance range of Fig. 6.16. Equation (6.8) can now be applied to the Mielenz maxima in Fig. 6.17(a) and (b) yielding corresponding relative SRE levels recorded in Table 6.3.

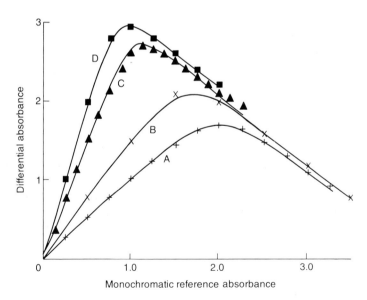

Fig. 6.18 [40]   Four plots of the differential absorbance versus the monochromatic reference absorbance of an arithmetic concentration series of Orleans Blue food-dye (E123) solutions placed in pairs of cells, the pathlength ratios of which were: A, 2.001; B, 2.499; C, 4.00; and D, 5.00. The measurements were made at 654 nm and with a spectral slitwidth of 1 nm.

Table 6.3 Relative SRE level in a Shimadzu 260 spectrophotometer set at 654 nm and a spectral slitwidth of 1 nm using the revised Mielenz et al. [4] and transmittance ratio test methods [40].

Cell pathlength ratio, α ($= b_s/b_r$)	2.001	2.499	4.00	5.00
Mielenz peak absorbance at 654 nm ($= -\log_{10} \rho$)	1.68	2.08	2.73	2.94
Monochromatic reference absorbance at 654 nm, A_λ	2.00	1.65	1.10	0.90
Relative SRE level from Equation (6.8) [$s(\times 10^{-4}) \pm 0.10$];				
(a) for $\nu = 1$	1.12	1.12	1.08	1.14
(b) for $\mu = t^A = 0.935^A$	1.53	1.51	1.46	1.55
Relative SRE level from Equation (6.3) and Fig. 6.18 maxima [$s(\times 10^{-4}) \pm 0.10$]	1.04	1.01	1.14	1.13
Relative SRE level from Equation (6.6) $t = 0.935$ and Fig. 6.18 maxima [$s(\times 10^{-4}) \pm 0.10$]	1.35	1.25	1.45	1.40

Equation (6.7) can now also be used to numerically generate four sets of data points, i.e. a matching set of data points for each set of experimental differential absorbance points in Fig. 6.18 via using trial s-values for the relative SRE level and setting $t = 0.935$. The ensuing simulated curves and experimental data points are traced in Fig. 6.17, and the optimum trial s-values are listed in the bottom row of Table 6.3.

The transmittance ratio test methods of Mielenz and Fleming yielded compatible relative SRE levels if the samples employed were assumed to be transparent to SRE, cf, rows 4 and 6 in Table 6.3. However, the direct transmittance test method of Fleming and O'Dea, which allows for the non-transparency of samples towards SRE, yielded a significantly higher relative SRE level of 0.00014 ±0.000015. The results from three test methods may be reconciled with one another within the experimental error if the sample's SRE transmittance is taken to be the value which the direct transmittance method yielded, i.e. 0.935.

6.8 Conclusion

Sharpe and Irish [31, 49] showed that virtually all SRE originates in grating imperfections, and that the remaining optics have a multiplicative rather than additive effect on the SRE. Hence, a cursory inspection of promotional literature received from instrument manufacturers through the years will indicate a gradual reduction in SRE errors due to the introduction of holographic gratings and front-coated mirrors. Hartree [16] speculated 'that the lower limit on the wavelength scale is not the one engraved by the manufacturer but the point below which stray-light errors rise above a tolerable level'. The lower UV wavelength may be affected by dissolved oxygen in samples and this may be remedied by degassing samples and flushing the instrument with dry N_2.

An idea for further reducing SRE may be to exploit the assumption that its spectral composition generally differs from that of the measuring wavelength and to employ detectors which are sensitive over a narrow spectral range. Diode array detection may facilitate this approach if cost considerations allow. SRE is usually specified at one or two wavelengths with the inference that the 'worst case' value is being represented. This may not be the case, but the more likely explanation is that the SRE has only been determined at the wavelength(s) specified for a representative sample of similar instruments. SRE data may not be used in a comparative sense [34], e.g. two instruments having the same relative SRE level while having different SRE spectral distributions. Since SRE test methods involve tedious procedures unsuited to routine testing, analysts, who tend to be too busy to appreciate the fine points of SRE measurement, are best advised to avoid highly absorbing samples (via lowering concentration and/or cell pathlength), especially if working with older instruments.

References

1. *Handbook for Unicam Spectrophotometer SP500 Series 2* (1968).
2. Edisbury, J.R. (1966) *Practical Hints on Absorption Spectrometry*, pp. 151–164. Adam Hilger, London.
3. Burgess, C. (1995) *Encyclopaedia of Analytical Science*, pp. 3643–3647. Academic Press, Adam Hilger.
4. Mielenz, K.D., Weidner, V.R. and Burke, R.W. (1982) *Applied Optics*, **21**, 3354.
5. *Handbook for Beckman Model DU Spectrophotometer*.
6. Beaven, G.H. (1949) *Photoelectr. Spectrom. Group Bulletin*, **1**, 7.
7. Goldring, L.S. (1950) *Photoelectr. Spectrom. Group Bulletin*, **2**, 34.
8. Perry, J.W. (1950) *Photoelectr. Spectrom. Group Bulletin*, **3**, 40.
9. Donaldson, R. (1950) *Photoelectr. Spectrom. Group Bulletin*, **3**, 45.
10. Martin, A.E. (1950) *Photoelectr. Spectrom. Group Bulletin*, **3**, 48.
11. Holliday, E.R. and Beaven, G.H. (1950) *Photoelectr. Spectrom. Group Bulletin*, **3**, 53.
12. Buist, J.G. (1981) *Techniques in Visible and Ultraviolet Spectrometry, Vol. 1, Standards in Absorption Spectrometry* (eds C. Burgess, and A. Knowles, pp. 94–110. Chapman and Hall, London.)
13. Goldring, S.L., Hawes, R.C., Hare, G.M., Beckman, A.O. and Stickney, N.E. (1953) *Anal. Chem.*, **25**, 869.

14. Lothian, G.F. (1963) *Analyst*, **88**, 678.
15. Menzies, A.C. (1960) *Pure Appl. Chem.*, **1**, 147.
16. Hartree, E.F. (1963) *Photoelecr. Spectrom. Group Bull.*, **15**, 398.
17. Rand, R.N. (1969) *Clin. Chem.*, **15**, 839.
18. Cook, R.B. and Jankow, R. (1972) *J. Chem. Educ.*, **49**, 405.
19. Beeler, M.F. and Lancaster, R.G. (1975) *Amer. J. Clin. Path.*, **63**, 953.
20. ASTM standard E-387 (*Annual Book of ASTM Standards*) (1984).
21. Sharpe, M.R. (1984) *Anal. Chem.*, **56**, 339A.
22. Francis, R.J. (1980) *International Laboratory*, **10** (May/June), 85.
23. West, M.A. and Kemp, D.R. (1976) *International Laboratory*, **6** (May/June), 27.
24. Irish, D. and Brickell, W.S. (1984) *Techniques in Visible and Ultraviolet Spectrometry, Vol. 3, Practical Absorption Spectrometry* (eds A. Knowles, and C. Burgess), pp. 48–65, 215–217. Chapman and Hall, London.
25. Fleming, P. (1990) *Analyst*, **115**, 1487.
26. Fleming, P. (1990) *Analyst*, **115**, 1577.
27. Tripp, T.P. and McFarlane, R.A. (1994) *Appl. Spectrosc.*, **48**, 1138.
28. Paul, J.B. and Saykally, R.J. (1997) *Anal. Chem.*, **69**, 287A.
29. O'Keefe, A. and Deacon, G.A.B. (1988) *Rev. Sci. Instrum.*, **59**, 2544.
30. Slavin, W. (1963) *Anal. Chem.*, **35**, 561.
31. Sharpe, M.R. and Irish, D. (1978) *Optica Acta*, **25**, 861.
32. Altmose, I.R. (1986) *J. Chem. Ed.*, **63**, A216.
33. Bausch and Lomb (1970) *Diffraction Grating Handbook*. New York.
34. Kaye, W. (1981) *Anal. Chem.*, **53**, 2201.
35. Poulson, R.E. (1964) *Applied Optics*, **3**, 99.
36. Zhu, D. and Qian, J. (1991) *Huaxue Tongbao*, **43**, 6.
37. Zhu, D. and Qian, J. (1992) *Chem. Abstrs*, **116**, 213670U.
38. Fleming, P. and O'Dea, J. (1991) *Analyst*, **116**, 195.
39. Fleming, P. (1990) *Analyst*, **115**, 375.
40. Fleming, P. (1991) *Analyst*, **116**, 909.
41. Pritchard, B.S. (1955) *J. Opt. Soc. Amer.*, **45**, 356.
42. Tarrant, A.W.S. (1978) *Optica Acta*, **25**, 1167.
43. Fleming, P. (1990) *Appl. Spectrosc.*, **44**, 522.
44. Neal, W.T.L. (1956) *Photoelectr. Spectrom. Group Bull.*, **9**, 204.
45. Neal, W.T.L. (1954) *Analyst*, **79**, 403.
46. Fleming, P. and Fleming, A. (1997) *Analyst*, **122**, 39.
47. Mavrodineanu, R. and Burke, R.W. (1987) *Advances in Standard Methodology Spectrophotometry* (eds C. Burgess and K.D. Mielenz), pp. 125–174. Elsevier Science, Amsterdam.
48. Gridgemann, N.T. (1952) *Anal. Chem.*, **24**, 445.
49. Sharpe, M.R. and Irish, D. (1976) *U.V. Spectrometry Group Bull.*, **4**, 51.

7 Wavelength Calibration

7.1 Introduction

All spectrometers should be checked regularly for wavelength accuracy for, despite the claims of manufacturers, prism and grating mountings and the driving mechanisms are susceptible to dirt, vibration and the effects of thermal expansion.

Many standards have been used in the past, each having its own particular merits, but few have withstood the test of time. The successful methods are still in use because of their simplicity and general reliability. A good standard should:

(a) be readily available;
(b) be easy to assemble or prepare;
(c) be easy and safe to handle;
(d) suit the optical properties of the instrument, e.g. spectral bandswidth, etc.;
(e) be unaffected by environmental conditions, e.g. have a small temperature coefficient, good chemical stability, etc.

The recommended methods fall into two categories: (i) the measurement of atomic emission lines from vapour discharge lamps; and (ii) the location of maxima in the absorption spectra of glass filters, solutions or crystals with very narrow absorption bands.

7.2 Line source standards

The most general and accurate calibration method is to introduce a discharge lamp into the lamp housing of the spectrometer. Table 7.1 lists some elements that have useful emission lines in the range 185–800 nm. Precise values for the location of these lines can be obtained from reference works, but it should be borne in mind that these have usually been measured *in vacuo*, and the lines from a normal lamp can vary by up to 0.1 nm from these values.

At shorter wavelengths, the mercury lines at 253.65 and 184.96 nm from a low-pressure lamp are particularly useful. At the other extreme, lithium

Table 7.1 The useful spectral ranges of selected vapour discharge lamps for wavelength calibrations. Precise values can be obtained from Ref. [17].

Element	Useful wavelength range (nm)
Mercury Zinc Cadmium	185–400
Magnesium Zinc Lead Cadmium Copper Mercury Rhodium	2003–420
Neon Tungsten Calcium Strontium Lithium Potassium Sodium	400–800
Krypton	up to 892.9
Argon	up to 811.5

lines at 670.8 and 610.4 nm, and potassium lines at 770.0 and 766.5 nm are conveniently placed. Neon has been recommended for the visible region [1], and Table 7.2 sets out some of these lines. The most-used region, between 250 and 580 nm, can be covered by a medium-pressure mercury lamp. Table 7.3 lists those bands which have been found by experience to be most readily detectable, sharp and free from shoulders.

Apart from these special sources, the deuterium emission lines in the visible region at 486.00 and 656.10 nm [2] are particularly useful for a quick check, since most spectrometers have a deuterium lamp as a source. The spectrum of the deuterium emission lines is shown in Fig. 7.1

7.3 Absorption standards

A more practical method for calibrating the wavelength scale of a spectrophotometer is the use of absorption wavelength standards. These can be glasses, aqueous solutions and crystals.

Holmium and didymium (a mixture of neodymium and praesodymium) oxide-containing glasses have been used extensively as absorption wave-

Table 7.2 The positions of the principal neon emission lines in air, from Ref. [1].

Wavelength (nm)
533.1
534.1
540.1
585.3
594.5
614.3
633.4
640.2
667.8
693.0
717.4
724.5

length standards over the past three decades. However, uncalibrated glass standards have now been shown to be rather unreliable as instrument quality improves, for the manufacture of reproducible glasses is extremely difficult. Their absorption bands, especially those of didymium, are relatively broad and their positions are sensitive to the slitwidth. Until recently, NIST supplied didymium glass filters as a wavelength standard, and Venable and Eckerle [3] list 15 absorption maxima ranging from 402 to 784 nm that are used as calibration wavelengths. Procedures for the location of the maxima are given, and the effects of SSW and temperature upon the apparent λ_{max} values were described in detail. Didymium glass filters have some useful bands in the near-UV, but usually become opaque

Table 7.3 Wavelength of selected absorption maxima for holmium and didymium glass filters [21]. Note that considerable variations are found between different batches of glass.

Wavelength of maximum (nm)	
Holmium glass	Didymium glass
241.5 ± 0.2	573.0 ± 3.0
279.4 ± 0.3	586.0 ± 3.0
287.5 ± 0.4	685.0 ± 4.5
333.7 ± 0.6	
360.9 ± 0.8	
418.4 ± 1.1	
453.2 ± 1.4	
536.2 ± 2.3	
637.5 ± 3.8	

Wavelength Calibration

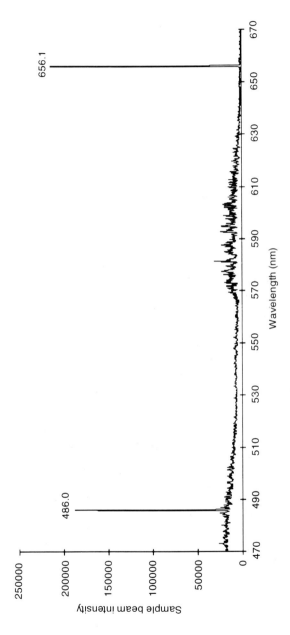

Fig. 7.1 Deuterium lamp emission spectrum. Courtesy of Unicam.

at shorter wavelengths. These standards are currently unavailable from NIST, but there are plans to reintroduce them to their product line. Selected maxima for typical holmium and didymium glass filters are given in Table 7.3: other values can be found in Refs [3–5]. The filters can only be recommended for wide-band instruments (SSW > 2 nm) and the uncertainties over the band positions should always be borne in mind. Both NIST and NPL provide calibrated holmium glass standards. Typical spectra for holmium and dydimium glasses are shown in Figs 7.2 and 7.3.

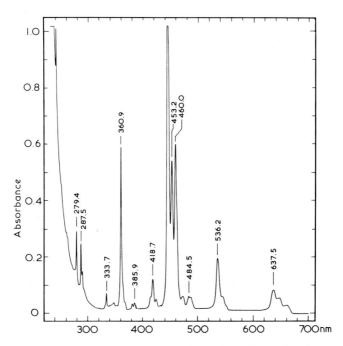

Fig. 7.2 Absorption spectrum of a holmium glass filter of unknown origin, scanned at 1 nm s^{-1} with SSW = 0.2. The peak values are taken from McNierney and Salvin [27]. Note that there are variations in peak positions between different batches of glass.

Recently, solid glass wavelength standards have become popular as internal calibration standards for UV spectrophotometers. A filter is built into the instrument and can be placed in the sample beam when required. The bands of the filter are checked when the instrument is turned on and there is also an option to perform a wavelength check when required.

An aqueous solution of rare-earth ions has narrower absorption bands than the glasses [6]: an absorption spectrum of holmium oxide in perchloric acid is given in Fig. 9.5. Tables 7.4 and 7.5 set out the principal lines of holmium (III) and samarium (III) ions measured in various laboratories. Within the temperature range 15–35°C, both glasses and

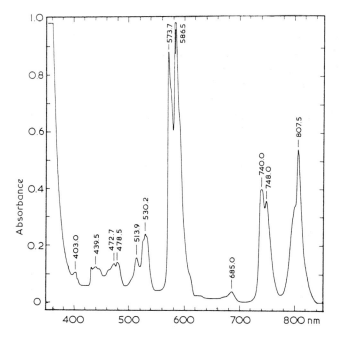

Fig. 7.3 Absorption spectrum of a didymium glass filter (Chance–Pilkington ON 12, 2.0 mm thick), scanned at 1 nm s^{-1} with SSW = 0.2. The peak values are taken from Perkampus [28]. Note that there are variations in peak positions between different batches of glass.

solutions of the rare-earth ions show little temperature variation, and this will only affect the second decimal place of the wavelength measured in nm [2]. A more important consideration is the SSW. A detailed study of some samarium and holmium lines as a function of bandwidth is presented in Table 7.6. Generally, an SSW that greatly exceeds the natural bandwidth of the line being measured will lead to a wavelength shift from the true value. Therefore, the proper use of rare-earth ion solutions, or indeed emission lines from discharge lamps, is restricted to narrow-band instruments, i.e. of SSW less than 5 nm.

NIST supply cuvettes of holmium oxide solution in perchloric acid as Standard Reference Material 2034 with a calibrated certificate giving λ_{max} values for 14 bands between 240 and 640 nm [7]. LGC in the UK also provide holmium oxide solution as a certified reference material with 16 bands between 219 and 865 nm.

Dielectric crystal wavelength standards have narrower absorption bands and are recommended for the wavelength calibration of instruments with SSW in the range 0.5–5 nm. An added advantage of crystal standards is their wider spectral transparency region extending into the UV and NIR. The McCrone crystal standard has many absorption bands across the entire UV,

Table 7.4 Wavelengths of the absorption maxima of holmium (III) ions in perchloric acid reported by various workers: I. Milazzo, Cary 17 [24], II. Vinter, Beckman Acta CV [22], III. Burgess, Cary 118 [25]; IV. Burgess, Perkin-Elmer 200 [26].

Wavelength of maximum (nm)			
I*	II†	III†	IV†
241.15	241.0	241.1	241.1
249.75	250.0	249.7	249.7
278.2	277.8	278.7	278.2
287.15	287.5	287.1	287.2
333.5	333.3	333.4	333.3
345.6	345.5	345.5	345.0
361.5	361.0	361.5	361.2
385.6	385.6	385.5	385.6‡
416.2	416.0	416.3	416.6‡
450.7	450.4	450.8	451.0‡
452.0			
467.75			468.0
485.25	485.2	485.8	485.2‡
536.3			536.8
640.5			

* The solution was 4 g of holmium oxide in 100 g 1.4 M perchloric acid
† The solution was 10 g of holmium oxide in 100 ml 17.5% w/v aqueous perchloric acid.
‡ These are complex bands

visible and NIR wavelength range [8]. NPL supplies these crystals in a frame which fits a 10 mm cuvette holder, and provides calibrated wavelength values for up to nine bands at three bandwidths. The nominal wavelength values for the McCrone crystal are given in Table 7.7. These values have been selected as the most appropriate for wavelength calibration as they are both narrow and symmetrical. Like all absorption standards, the positions of the absorption maxima are sensitive to SSW. Typical spectra from the McCrone crystal are shown in Figs 9.8 and 9.9.

Two Russian crystal wavelength standards are also available from the NPL extending wavelength coverage in the UV and NIR. These are supplied to NPL from the Vavilov State Optical Institute in St. Petersburg. Together with the McCrone standard, they provide bands from 208 to 1735 nm. NIST are also developing further wavelength standards for the NIR region.

For far-UV calibration, oxygen and iodine vapour have absorption lines that are well characterized [7]. Iodine solutions have absorption maxima at 174.2, 174.8, 175.5 and 176.2 nm [8], but these solutions must be ultraclean and free from oxygen [9].

Table 7.5 Wavelengths of the absorption maxima of samarium (III) ions in perchloric acid reported by various workers: I. Vinter, Beckman Acta CV [22]; II. Burgess, Cary 118 [25]; III. Burgess, Cary 219 [26]; IV. Burgess, Perkin-Elmer 200 [26].

Wavelength of maximum (nm)			
I	II	III	IV
305.4	304.8	305.2	304.8
317.5	316.6	317.2	317.1
332.2	331.2	331.9	331.6
344.8	344.2	344.2	344.1
362.7	361.9	362.2	362.1
374.5	374.1	374.0	374.1
391.1	389.8	390.4	390.1*
	400.9	401.0	401.2
407.5	406.4	406.6	406.7*
415.8	414.6	415.0	415.0*
417.7	416.6	417.0	416.8*
442.3	441.8	441.0	441.0
464.2	463.1	463.1	463.0*
479.6	478.4	478.5	478.9*
500.2	498.5	499.0	498.9*

The solution was 10 g samarium (III) oxide in 100 ml 17.5% w/v aqueous perchloric acid
* Multiple bands

The methods given in this and the preceding section are recommended as good wavelength standards. In the next section, more specialized procedures are outlined, and the excellent review article by Alman and Billmeyer [12] will be found to be a useful piece of supplementary reading.

7.4 Other methods

7.4.1 The Van den Akker method for wide-band instruments

For instruments of spectral slitwidth greater than 10 nm and for colorimeters, Van den Akker suggested a technique using a carefully chosen glass filter [13]. A mercury vapour emission line in the appropriate spectral region is selected and a filter is chosen whose transmission curve is approximately linear in this region, i.e. the transmittance over the bandpass λ_1 to λ_2 is a linear function of wavelength. This will not be true near peaks or troughs in the transmission curve. If the apparent transmittance of the filter is measured with an SSW of $\lambda_1 - \lambda_2$, then there will be a wavelength λ_c at which the same transmittance value would be seen if

Table 7.6 The variation of the apparent λ_{max} values with slitwidth for narrow-band spectra. Solutions of holmium (III) oxide and samarium (III) oxide in perchloric acid measured on a Beckman Acta CV (linear wavenumber) instrument using various programmed slit settings [22].

		Holmium bands			Samarium bands						
Nominal λ_{max} (nm)		361.0	245.5	333.3	401.0	374.5	362.7	344.8	332.2	317.5	305.4
NBW of band (nm)		4.2	3.6	3.3	—	4.2	3.9	4.1	3.3	4.1	3.7
Slit programme:											
I	SSW	0.1	0.1	0.1	—	0.1	0.1	0.2	0.2	0.2	0.2
	λ	361.3	345.7	333.6	—	374.9	362.65	344.8	332.2	317.7	305.8
II	SSW	0.1	0.2	0.2	0.3	0.3	0.35	0.45	0.6	0.7	0.7
	λ	361.3	345.4	333.4	401.1	374.2	362.3	344.5	331.95	317.5	305.4
III	SSW	0.3	0.35	0.4	0.5	0.6	0.8	1.1	1.5	1.6	1.5
	λ	361.3	345.4	333.5	401.2	374.7	362.7	344.6	332.0	317.7	305.5
IV	SSW	0.5	0.7	0.9	—	1.2	1.4	1.7	2.3	2.7	2.5
	λ	361.3	345.4	333.5	—	374.5	362.7	344.8	333.9	317.5	305.4
V	SSW	0.8	1.0	1.3	1.3	1.7	2.0	2.9	3.5	4.1	3.9
	λ	361.3	345.4	333.4	401.2	374.25	362.3	344.3	331.95	317.2	305.2
VI	SSW	1.0	1.4	1.7	2.7	3.5	4.1	5.8	7.0	8.2	8.2
	λ	361.3	345.3	333.3	401.2	373.8	361.8	343.6	331.7	316.45	303.95
VII	SSW	1.4	1.7	2.2	—	—	—	—	—	—	—
	λ	361.3	345.3	333.5	—	—	—	—	—	—	—
VIII	SSW	2.0	2.6	2.3	—	—	—	—	—	—	—
	λ	361.2	345.4	333.6	—	—	—	—	—	—	—
IX	SSW	4.1	5.2	6.4	—	—	—	—	—	—	—
	λ	361.3	345.4	333.3	—	—	—	—	—	—	—
X	SSW	8.1	8.1	8.1	—	—	—	—	—	—	—
	λ	361.0	345.1	329.8	—	—	—	—	—	—	—

NBW: natural bandwidth of band in nanometres
SSW: special slitwidth in nanometres

Table 7.7 Nominal wavelength values for the McCrone filter.

Wavelength (nm)	Region
253.7	UV
254.9	UV
353.9	Visible
481.3	Visible
588.6	Visible
869.2	NIR
1486.0	NIR
1735.2	NIR

the measurement was performed with a truly monochromatic source; λ_c will lie between λ_1 and λ_2 but will not necessarily be the mid-point.

The filter is measured using the mercury source in place of the normal lamp. The normal lamp is then placed in position and the wavelength scale scanned until the same transmission value is observed: the wavelength scale then gives the value for λ_c. Van den Akker found that this value of λ_c should fall within 0.8 nm of the true wavelength of the line source. The method overcomes slit parameter errors and is dependent for its accuracy mainly upon the slope of the filter transmission at the calibration point. A slope of 1% per nm represents a 0.1 nm uncertainty in the final value of λ_c.

7.4.2 Interference filters

Heidt and Bosley [14] have described the manufacture and use of an interference cell that consists of a pair of closely-spaced parallel plates that generates a transmission spectrum that is a series of maxima and minima. The cell is calibrated by using a line source in the same instrument. Any part of the spectrum can be covered by varying the separation of the plates. Since the interference maxima are rather broad, it is difficult to locate the peak exactly. The slit function must be symmetrical, and a temperature effect will be seen if the cell spacer is liable to expand.

Buist has pointed out that an ordinary infra-red cell equipped with fused silica or, better, calcium fluoride windows will give these fringes, and with a 25 µm spacer it is possible to generate fringes over the 220–850 nm range [15].

7.4.3 Retardation plates

By constructing a cell consisting of a Nicol prism polarizer, a quartz birefringent plate and a Nicol prism analyser, Buc and Steams were able to calibrate their spectrometer in a non-empirical way without recourse to calibration materials [16]. The method depends upon the fact that the quartz plate splits plane-polarized light into two orthogonal plane-polarized beams of differing velocity. Depending upon their phase relationship at the exit surface, a series of transmission maxima and minima are seen across a large portion of the UV–visible range. The positions of these maxima can be calculated from a knowledge of the cell dimensions and the birefringence of quartz. The temperature must also be known and, although the slit function should be symmetrical, its width and slope are irrelevant. The authors were also able to use this method to estimate slit shapes and widths.

7.4.4 The cross-filter technique

This method involves the use of two filters whose transmission curves cross at some point [17]. This point is determined on an instrument that has already been calibrated. The method is susceptible to slit variation, it is a one-point calibration and it would be necessary to determine whether the transmission curve of either filter changed with temperature.

7.5 Conclusion

For narrow-band instruments, i.e. of SSW less than 5 nm, line sources are the most effective means of wavelength calibration. When performing the calibration, as Robertson [18] has pointed out, peripheral equipment should also be checked for performance, e.g. the response speed of the data capture whether it be a computer or a pen recorder, for this can introduce errors into the calibration. The McCrone filter provides a very convenient wavelength calibration as it can be fitted into the sample holder.

For wide-band instruments, the rare-earth glasses are convenient to use. Van den Akker's procedure is of particular relevance when studying materials with very broad absorption bands, e.g. those encountered in colorimetry. The interference and retardation techniques are probably best left to specialists.

Calibrations using a mercury arc lamp, McCrone crystal and holmium perchlorate solutions are described in detail in Chapter 9.

References

1. Menzies, A.C. (1960) *Pure & Appl. Chem.*, **1**, 147.
2. Reule, A.G. (1976) *J. Res. Nat. Bur. Std*, **80A**, 609.
3. Venable, W.H. and Eckerle, K.L. (1979) NBS Special Publication No. 260-66.

4. McNeirney, J. and Slavin, W. (1962) *Appl. Opt.*, **1**, 365.
5. *UV Atlas of Organic Compounds* (eds H. Perkampus, I. Sandeman, and C.J. Timmons (1971), Vetlag Chemie.
6. West, M.A. and Kemp, D.R. (1976) *Int. Lab.*, May/June, 27.
7. Weidner, V.R. *et al.* (9186) NBS Special Publication 260–102.
8. Verrill, J.F. (1995) *Spectrometry, Luminescence and Colour*. Elsevier Science.
9. Kaye, W.I. (1961) *Appl. Spec.*, **15**, 89.
10. Cordes, H. (1935) *Z. Physik*, **97**, 603.
11. Bramston-Cook, R. and Erickson, J.O. (1973) *Vanan Instruments Applications*, **7**.
12. Aiman, D.H. and Billmeyer, F.W. (1975) *J. Chem. Educ.*, **52**, A315.
13. Van den Akker, J. (1943) *J. Opt. Soc. Amer.*, **33**, 257.
14. Heidt, L.J. and Bosley, D.E. (1953) *J. Opt. Soc. Amer.*, **43**, 760.
15. Buist, G.J. (1976) *J. Chem. Ed.*, **53**, 727.
16. Buc, G.L. and Stearns, E.I. (1945) *J. Opt. Soc. Amer.*, **35**, 465.
17. Sanghi, I. and Parthasarathy, N.V. (1959) *Naturwiss.*, **46**, 315.
18. Robertson, A.R. (1976) *J. Res. Nat. Bur. Std.*, **80A**, 625.
19. Harrison, G.R. (1969) *Wavelength Tables*. MIT Press, Cambridge, MA.
20. Zaidel, A.N. (1961) *Tables of Spectrum Lines*. p. 381. Pergamon Press.
21. Gibson, K.S. (1949) NBS Circular No. 484, pp. 13–15.
22. Vinter, E., Unpublished observations.
23. Edisbury, J.R. (1966) *Practical Hints on Absorption Spectrometry*. Hilger & Watts, London.
24. Milazzo, G. (1976) Subcommittee on Calibration and Test Materials. IUPAC commission on physicochemical measurements and standards.
25. Burgess, C. (1977) *UV Spec. Grp. Bull.*, **5**, 77.
26. Burgess, C. Unpublished observations.
27. McNierney, J. and Slavin, W. (1962) *Appl. Optics*, **1**, 365.
28. Perkampus, H. (1991) *UV Atlas of Organic Compounds*, 2nd ed, VCH.

8 Regulatory Overview

8.1 Introduction

Today, thanks to the genius of researchers, manufacturers and commerce, there are more products on sale than ever before in the history of mankind. Increasing complexity is leading to a greater need for consumer protection. More and more products require testing and calibration to ensure compliance with specification and safety regulations before release to the market.

The growth in international trade, particularly in Europe with its variety of traditions, languages, standards and habits, is adding to this need for consistency. Trade, even in simple commodities and products, demands supporting technical data, only available from calibrations and tests. The achievement of the Single European Market and the formation of the World Trade Organisation (WTO) require cross-frontier acceptance of test results, and as so many nationally accepted calibration and test methods differ, there must be an acceptable means of determining their conformity.

The enforcement of regulation and often the administration of justice in the fields of health, safety and environmental protection need to be based on a foundation of valid calibration and testing in which the public has confidence. Competent calibration and traceability of measurements is essential for industrial manufacture and is a major criterion in addressing product liability. It features prominently in certification to ISO 9000. It also underpins all testing. The credibility of test results depends on accuracy, precision, repeatability and reproducibility. In turn, these depend on the competence of the tester and the validity of the methods used. Bodies who have to accept goods must have confidence that laboratories conducting tests and calibrations are competent, and that their results are valid.

Generation of 'quality data' has always been one of the primary responsibilities of any laboratory manager; but in recent years with the introduction of several internationally recognized quality standards, this requirement has become an absolute necessity.

In many laboratories, the instrument is still perceived as the box of electronics, possibly combined with additional physical hardware; e.g. a spectrometer optical system and/or personal computer. However, in many current, and an ever increasing number of laboratory environments, this simple definition no longer even begins to describe the required interpretation of this entity as an instrument system. How the system has expanded, and continues to widen its boundaries is covered by specific reference to qualification, training, etc. These are presented as the so-called 'ethereal' considerations of an instrument system.

International accreditation and regulatory bodies expect that the instrument system and methods used to generate critical analytical data will be validated. Similarly, 'quality' data and methods can only be produced by using systems proven to be under control. For the above reasons, a given laboratory may be working to any one of a number of quality assurance standards, including their own internal quality assurance; however, the three most usually encountered are Good Laboratory Practice (GLP), ISO 9000 and ISO/IEC Guide 25 (ISO 17025).

8.2 Recent history

In the early 1970s, regulation in many countries in the field of pharmaceutical manufacturing initiated the production of guide(s) for Good Manufacturing Practice (GMP). As a follow up to these procedures, during the late 1970s these regulatory organizations introduced quality schemes, where clearly the main aim of the procedures put in place was to address this data quality issue. Who they are has already briefly been mentioned; how they interact will be discussed later on.

However, in 1992, a major international survey of analytical scientists, coupled with substantial evidence still indicated that whilst a significant improvement had been observed in many areas, analytical data were still often not fit for their intended purpose.

CITAC – Co-operation on International Traceability in Analytical Chemistry – arose out of an international workshop held in association with the Pittsburgh Conference in Atlanta in March 1993. The aim of this workshop was to discuss how analytical activities could be developed to meet the needs of the 21st century, and it identified a wide variety of issues to be addressed to ensure that analytical measurements made in different countries or at different times are comparable. These range from the development of traceable reference materials and methods to the harmonization of analytical quality practices.

The CITAC Initiative aims to foster collaboration between existing organizations to improve the international comparability of chemical measurement. A Working Group takes matters forward and its initial activities have centred on a few specific high-priority activities.

It is from this point on that manufacturers of UV–Visible instruments observed an acceleration in the response of organizations with functioning laboratories to this fundamental issue.

Generically, the catalyst of this acceleration had been, and continues to be, the requirement to meet the demands of external forces. Some specific examples of the many available are cited here.

Firstly, all regulatory organizations demand certain standards to be met, e.g. Good Manufacturing Practice (GMP) and Good Laboratory Practice (GLP) compliance are both essential for US Food and Drug Administration (FDA) approval of a drug submission.

Secondly, there is the reputation of the laboratory, both internal within a company, and external; contract laboratories rely on the proven generation of quality data for their very existence, a good reputation brings new business, a bad reputation can destroy everything.

To quote a speaker at a Pharmaceutical Conference in Edinburgh in 1993.

> ... The 1990s are the decade of quality. Laboratories must publish quality information or they will not exist.

Lastly, companies are looking to improve their overall quality, not just in the laboratory, but across the whole organization, to increase their competitiveness in the market place. This is often reflected by the appointment of a Quality Manager, responsible for the validation of systems or processes within a company, and/or the introduction of the ISO 9000 group of quality standards.

Therefore, for many reasons, a given laboratory may be working to any one of a number of quality assurance practices. As has already been stated, the three most usually encountered are Good Laboratory Practice (GLP), ISO 9000 and ISO/IEC Guide 25 (ISO 17025).

Reviewed below, these internationally recognized processes all recognized the fundamental requirement of calibration and traceability of measurement of an instrument system. Unfortunately, in some areas this is not matched by the availability of Certified Reference Materials (CRMs).

8.3 Good Laboratory Practice (GLP)

GLP is intended to promote the quality and validity of test data. It is a managerial concept covering the organizational processes and the conditions under which laboratory studies are planned, performed, monitored, recorded and reported.

Formally introduced in 1976 by the FDA for use in non-clinical studies, these regulations were used as a basis to construct the internationally recognized Organization for Economic Co-Operation and Development (OECD) guidelines, issued in 1981.

8.4 ISO 9000

ISO develops and maintains the ISO 9000 series of International Standards for quality management and quality assurance. First introduced in 1987, this series of standards has been employed by thousands of businesses worldwide and has been adopted by more than 70 countries as the national standard. ISO 9002 is probably the most usually encountered as it includes the quality system requirements for production, installation and servicing. However, it is worth remembering that '... you cannot test quality into a system, it must be designed and built in from the outset', and therefore using this premise, logically, generation of good quality data only comes from consideration of this fundamental requirement at the design stage. Of all the ISO 9000 series, only ISO 9001 includes the quality system requirement for product design.

These first two schemes result in the certification of a laboratory to the appropriate standard, and relate to the audited implementation of these quality management schemes. The last of these three internationally recognized schemes is somewhat different in its approach, and uses the term accreditation to describe the process.

8.5 ISO/IEC Guide 25 (ISO 17025)

Accreditation is the formal recognition of the competence of a body or organization for a well-defined purpose. In contrast with the GLP and ISO 9000 processes, accreditation of a laboratory to ISO/IEC Guide 25 (ISO 17025) involves assessment of the technical competence and capability of the laboratory and its personnel. In practice, it is the procedure by which a laboratory is assessed to perform a specific range of tests or measurements. Specific areas examined include infrastructure and staff qualifications; in addition to checks that an adequate quality management scheme is in place. The accreditation covers the range of materials tested or analysed, the tests carried out, the method and equipment used, and the accuracy or precision expected, and is specific to the facility and the test.

Most national laboratory accreditation schemes have based their standards on the international document ISO/IEC Guide 25, which establishes the process not only as a third party audit, but an assessment of the data generation process by a 'peer review'.

In practice, the ISO/IEC Guide 25 is used at a national level to construct the corresponding national accreditation standards. This standard is currently implemented in the UK by the United Kingdom Accreditation Service (UKAS).

How these accreditation bodies have evolved and developed can be shown by detailing the chronological changes to 1997 of two bodies in this area; one national, the other international.

8.6 International chronology – International Laboratory Accreditation Conference (ILAC)

ILAC was originally initiated in Copenhagen in 1977 when the first international meeting was held to discuss laboratory accreditation and its potential to support trade through reduction of technical barriers to trade. Known until 1996 as the International Laboratory Accreditation Conference (ILAC), it has met 15 times around the world up to and including the 1998 Conference in Sydney.

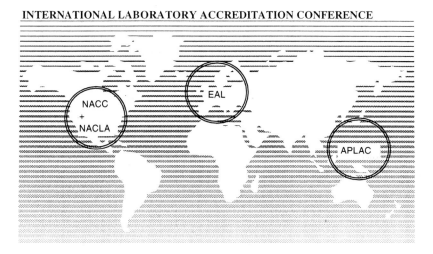

Fig. 8.1 Distribution of regional fora.

The growth of laboratory accreditation worldwide has motivated the creation of regional fora for cooperation and development. To date, there are two formal regional laboratory accreditation cooperations covering both calibration and testing, and one covering calibration only. The formation in 1997 of the National Council for Laboratory Accreditation (NACLA) in the USA, a co-ordination body, similar in structure to the established regional fora, must also be documented at this point. These are:

APLAC – Asia Pacific Laboratory Accreditation Cooperation
Started in 1992 – MOU signed by 16 countries in 1995.

EAL – European Cooperation for Accreditation of Laboratories
Formed in 1994 – 17 countries participating. Became EA – European Cooperation for Accreditation – in 1998.

NACC – North American Calibration Cooperation
MOU signed in 1994 between Canada, Mexico and USA.

8.7 National chronology – United Kingdom Accreditation Service (UKAS)

Formed in August 1995 by the amalgamation of the National Measurement Accreditation Service (NAMAS) and the National Accreditation Council for Certification Bodies (NACCB). NAMAS (now renamed National Accreditation of Measurement and Sampling) has been active as the UK accreditation body since 1985. In June 1994, NAMAS and its partners in the Western European Laboratory Accreditation Cooperation (WELAC), combined with the Western European Calibration Cooperation (WECC) to form the European Cooperation for Accreditation of Laboratories (EAL). This harmonization to date has resulted in the following list of countries participating in bilateral and/or multilateral agreements with UKAS: Australia, Denmark, Finland, France, Germany, Hong Kong, Ireland, Italy, The Netherlands, New Zealand, Norway, South Africa, Spain, Sweden and Switzerland. Currently, the following countries are signatories to the Memorandum of Understanding (MOU) that established EAL, but have yet to be evaluated and admitted to the mutual recognition agreements: Austria, Belgium, Greece, Iceland and Portugal.

If one looks at the chronological sequence, clearly the formation of UKAS was instigated by the combination of the 'accreditation' and 'calibration' bodies at the European level.

The main advantage of this cooperation to an accredited laboratory is the unquestioned acceptance of test reports and calibration certificates throughout the member states.

8.8 ISO 9000 versus ISO/IEC Guide 25 (ISO 17025)

It is not inconceivable that there will be occurrences when a chemical testing laboratory may be asked to comply to two systems; e.g. consider a laboratory seeking accreditation to ISO Guide 25 (for the reasons described above) within a company seeking overall ISO 9000 certification. This potential for the duplication of work in common areas was again one of the reasons for the formation of CITAC.

The difference between ISO 9000 certification and ISO/IEC Guide 25 accreditation may be summarized thus:

Quality system registration (ISO 9000) asks:

- Have you defined your procedures?
- Are they documented?
- Are you following them?

Laboratory accreditation asks the same questions, but then goes on to ask:

- Are they the most appropriate test procedures to use in the circumstances?
- Will they produce accurate results?
- How have you validated the procedures to ensure their accuracy?
- Do you have effective quality control procedures to ensure ongoing accuracy?
- Do you understand the science behind the test procedures?
- Do you know the limitations of the procedures?
- Can you foresee and cope with any technical problems that may arise while using the procedures?
- Do you have all the correct equipment, consumables and other resources necessary to perform these procedures?

The registration of a laboratory's quality management system is a component of laboratory accreditation, not a substitute. Quality system registration of a laboratory to ISO 9000 misses a key element, technical validity and competence.

This 'definition' of ISO 9000 as a quality management subset is reflected in the draft revision of ISO/IEC Guide 25, and also on certificates currently being issued by national accreditation bodies.

The extract below is taken directly from a certificate issued by UKAS.

> ... Accredited organisations meet the requirements of EN 45001, ISO/IEC Guide 25 and the relevant requirements of the BS EN ISO 9000 series of standards, including those of the model described in BS EN ISO 9002 when acting as suppliers producing calibration results.

Clearly, the 'Holy Grail' would be one definitive quality scheme, internationally recognized and capable of fulfilling all the certification and accreditation processes. Whilst this may be some way off, there are some significant recent moves on the path to the process of harmonization.

In December 1995, CITAC issued the first edition of Guide 1 'International Guide to Quality in Analytical Chemistry – An Aid to Accreditation'.

This document aims to provide laboratories with guidance on the best practice for improving the quality of the analytical operations they carry out. For the first time in any documentation relating to quality schemes, where applicable, each section is cross-referenced to the related parts of ISO Guide 25, ISO 9000 and OECD GLP principles.

In 1998, CITAC issued their second guide 'QA Best Practice for Research and Development and Non-routine Analysis'.

This document is a guidance document on the quality system requirements for non-routine and R&D activities. The present interpretation of quality system standards by the national accreditation authorities tends to be overly prescriptive and generally applicable only to specific types of

8.9 Qualification and 'ethereal considerations'

Instrumentation will only be fit for its purpose if it is properly calibrated before use. In some cases (e.g. for the wavelength scale of a spectrometer) calibration may form part of the equipment qualification process, which is

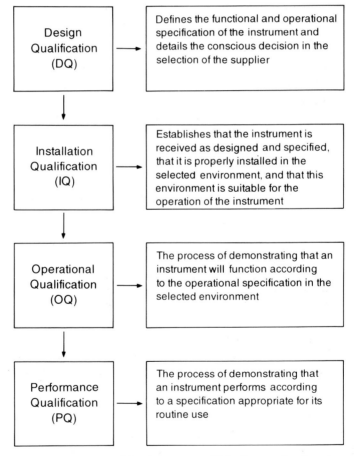

Fig. 8.2 The equipment qualification process. This diagram is taken from a guidance document published by a Eurachem Instrumentation Working Group, where a more complete description of the process is available.

well established and proven in the highly regulated pharmaceutical industry. Equipment qualification is a systematic procedure which ensures that suitable equipment is purchased and that it remains fit for its chosen purpose throughout its operating life.

This qualification is one of the most important parts of any customer's validation plan with procedures put in place to ensure the instrument system is initially in calibration, and remains so for the rest of its working life. However, the primary calibration obviously forms part of the manufacturing process as a vendor responsibility.

The responsibility of this process resides at the beginning with the vendor, and passes through a stage of mutual control, until ultimately it is passed to the user.

Calibration is at the centre of the whole quality management of a laboratory, and as we have seen inputs into both the qualification and validation requirements for a system.

Figure 8.3 shows this 'layered process' which impacts on both the user and vendor. The fundamental difference is that whilst the user defines an overall TQM/validation plan for the laboratory, working inward through the qualification of the systems, to calibration; the vendor works outwards from the initial calibration at manufacture, assisting the user in the

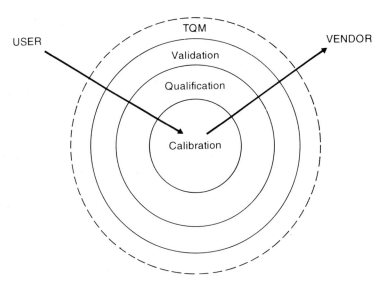

As a user:
1. Establish TQM protocols
2. Formulate validation plan
3. Qualify instrument system
4. Ensure initial (and maintain) calibration

As a vendor:
1. Ensure calibration to specification
2. Assist user in the qualification at the system location
3. Assist/advise on additional validation/ training/TQM aspects

Fig. 8.3 'VQC target'.

qualification of the systems, training, etc., the so called 'ethereal considerations'.

8.10 What is the implication for the lab manager of the year 2000?

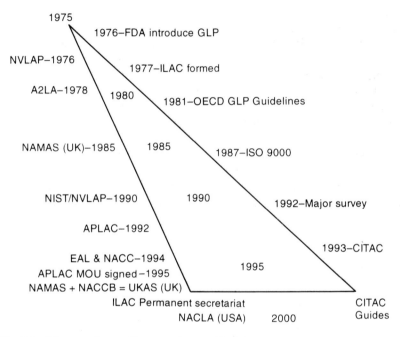

Fig. 8.4 History of compliance and accreditation.

To a laboratory manager, 'quality consciousness' will become one of the most important considerations of the whole laboratory environment, and to us as instrument vendors this clearly has an impact when the purchase of new instrument systems is being considered. To modify the previously used quotation slightly.

> ... In the new Millennium, instrument vendors must be capable of assisting the production of quality data and the proving that it is fit for purpose, or they will cease to exist.

An increasing number of users of instrument systems will have to formally prove that a system is or has been under control, generating quality data that are fit for the purpose; and these customers will purchase from instrument vendors providing the tools to assist them with this burden of proof.

9 Recommended Procedures for Standardization

Standards are the spectroscopist's 'links with sanity'. The ideal profile for standards has been discussed by Verrill [1]. The requirements for photometric accuracy, wavelength accuracy and stray-light standards are represented ideally by perfect neutral density filters, sharp isolated absorption peaks and a perfect cut-off filter. They are illustrated in Fig. 9.1

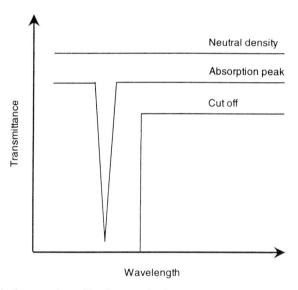

Fig. 9.1 Ideal spectral profiles for standards.

However, these are not realized in practice and, moreover, are not always invariant with time or temperature. This chapter is about a selection of standards and procedures which currently reflects best practice.

9.1 Resolution of monochromators

The best performance of a spectrometer will only be attained – in terms of both absorbance and wavelength accuracy – if careful consideration is given to the resolution of the monochromator. Since resolution is a function of slitwidth as well as dispersion of the instrument, the choice of slit setting is a critical one. Most modern instruments use grating monochromators which provide constant dispersion with wavelength. The smaller the spectral bandwidth, the greater the resolution, but the corresponding reduction in energy means that the signal-to-noise ratio falls. It is therefore necessary to select the smallest possible slitwidth that gives an acceptable noise level. When measuring an absorbance band in a high-resolution instrument, it is recommended that the spectral band width (SBW) should not exceed 10% of NBW of the band. There are a number of simple checks for the resolution of an instrument.

(a) Record the spectrum of a solution of a 0.02% v/v solution of toluene in hexane compared with a solvent blank. The ratio of the maximum at 269 nm and the minimum at 266 nm gives a measure of the resolution of the instrument.

The ratio values are within ± 0.1 for temperatures between 15 and 30°C, and concentrations of toluene between 0.005 and 0.04% v/v [3]. The European Pharmacopoeia specifies a limit for resolution of not < 1.5. This equates to a SBW of 1.8 nm. Typical spectra are shown in Fig. 9.2.

(b) Measure the profile of the 656.1 nm line from the emission of the instrument's deuterium lamp using the single-beam or 'energy' mode of the instrument. Figure 9.3 shows some typical profiles. The apparent width of the band at half-peak height is taken to be the SBW of the instrument [4]. This is illustrated for a SBW of 0.2 nm in Fig. 9.4. Note that this approximation assumes an isolated band. The 656.1 nm deuterium line has a closely associated hydrogen line. For the most accurate work, the use of a laser line, typically from a low-power helium–neon laser, is preferred. NIST employ the 435.83 nm mercury line for determining spectral band width [5].

(c) For an indication of spectral bandwidths less than 0.5 nm, the resolution of minor peaks in the benzene vapour spectrum in the region

Table 9.1 Variation of toluene band ratios with SBW [2].

Spectral bandwidth (nm)	Observed ratio [2]
0.25	2.3
0.5	2.2
1.0	2.0
2.0	1.4
3.0	1.1
4.0	1.0

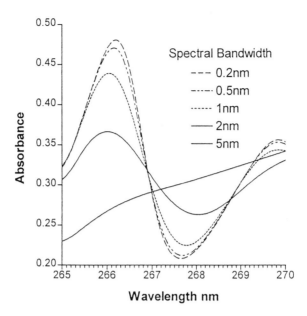

Fig. 9.2 Variation of spectrum of 0.02%v/v toluene in hexane at 25°C with spectral bandwidth. Jasco V560 spectrometer, scan rate 10 nm min^{-1} and data pitch 0.025 nm.

of 260 nm is a very useful guide. The use of benzene has been restricted in recent years and must be used with care. However, sealed cuvettes containing benzene vapour are commercially available for this purpose. For those able to carry out the safe handling of benzene, one drop of the liquid in a 10 mm stoppered silica cuvette will provide a saturated vapour at room temperature. This will give a maximum absorbance value of about 1.6 for a SBW of 0.1 nm. Typical spectra for bandwidths of 0.1, 0.2, 0.5, 1 and 2 nm are shown in Fig. 9.5. The spectra, corrected for cuvette contribution, have been off-set by 0.75, 0.15, –0.17, –0.5 and –0.7 absorbance units for display purposes. The values for the peak maxim are indicated by Lang [6]. The 253.49 and 259.56 nm peaks are only visible if the SBW is less than 0.2 nm. Above 1 nm all the fine structure is lost.

9.2 Wavelength calibration

9.2.1 High-accuracy calibration

A discharge lamp is recommended for this purpose. A low-pressure mercury lamp has a number of intense lines that cover a large part of the UV and visible range. The lamp should be placed as near as possible to the

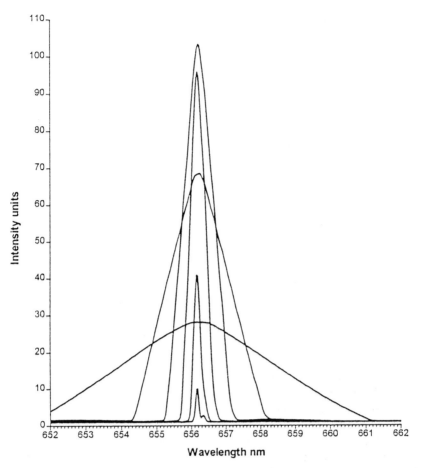

Fig. 9.3 Variation of the 656.1 nm deuterium line profile at 25°C with spectral bandwidth between 0.1 and 5 nm. Jasco V560 spectrometer, single-beam mode, scan rate 10 nm min^{-1}, fast detector response and data pitch 0.025 nm.

entry slit of the monochromator. 'Pen-ray' discharge lamps are well suited to this purpose, particularly mercury. The instrument should be operated in single-beam or 'energy' mode, with slitwidths as small as practicable.

With recording instruments, the slowest scanning speed should be used. All measurements should be made with the instrument scanning in the normal direction to avoid the effects of backlash. Since temperature affects the wavelength setting of most instruments, the instrument should be allowed to warm up to its normal operating temperature, and the ambient temperature noted. The most useful lines, Hg lines, in air [7], are 253.65, 289.36, 296.73, 365.02, 366.33, 404.66, 435.83, 546.07, 576.96 and 579.07 nm. A typical spectrum is shown in Fig. 9.6.

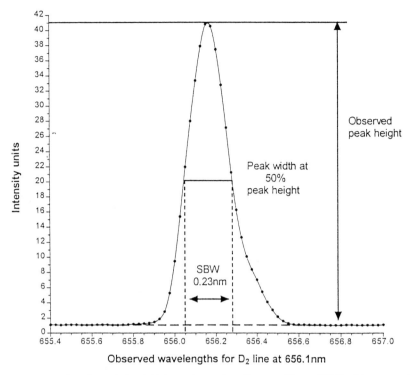

Fig. 9.4 Estimation of the 656.1 nm deuterium line profile at 25°C with spectral bandwidth set at 0.2 nmm. Jasco V560 spectrometer, single-beam mode, scan rate 10 nm min^{-1}, fast detector response and data pitch 0.025 nm.

9.2.2 Routine calibration

(a) The peak positions of a solution of holmium (III) ions form a convenient standard that can be read to the nearest 1 nm with ease. Solutions of 5% w/v holmium oxide in 10% v/v perchloric acid are recommended and can be obtained commercially. Scan a 10 mm pathlength of the solution at $25 \pm 5°C$ at low speed (40 nm min^{-1} is suitable) with a spectral bandwidth of 1 nm. A typical spectrum is given in Fig. 9.8. Locate the following 12 prominent bands, given to the nearest 0.1 nm: 241.1, 250.0, 278.1, 287.2, 333.4, 345.5, 361.3, 385.7, 416.3, 451.3, 467.8, 485.3 nm [5]. Two more bands at 536.7 and 640.5 nm are available in the visible. Note that the maxima move slightly with SBW, and that for SBW values less than 0.5 nm the band at 451.3 nm is resolved into two bands. However, these values are adequate for most solution work for ensuring wavelength calibration to ± 1 nm.

(b) Alternatively, the McCrone filter described in Chapter 7 can be used for wavelength calibration. The six recommended bands in the UV/

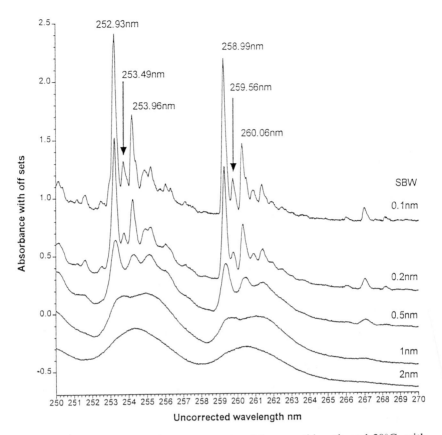

Fig. 9.5 Benzene saturated vapour spectra, 10 mm pathlength and 20°C, with spectral bandwidths between 0.1 and 2 nm. Jasco V560 spectrometer, scan rate 10 nm min^{-1}, quick detector response and data pitch 0.025 nm.

visible region are 253.6, 254.9, 353.9, 481.3, 588.6 and 748.4 nm, and the three in the NIR are 869.2, 1486.0 and 1735.2 nm. The uncertainties of these bands are ±0.1 nm for the UV/visible region and ±0.4 nm for the NIR for NPL calibrated filters. Typical spectra are shown in Figs 9.8 and 9.9.

9.3 Stray-light measurement

A complete analysis of the stray-light characteristics of an instrument requires the measurement of monochromator stray-light, but since this is a tedious process requiring specialized equipment, it is not usually undertaken. Instrumental stray-light (ISL) is more significant as far as stray-light errors are concerned, and depends not only on monochromator

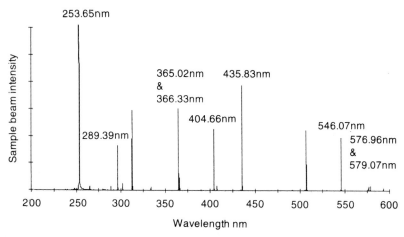

Fig. 9.6 Typical mercury 'Pen Ray' lamp emission spectrum [8].

stray-light but the absorbance of the sample being measured and the response of the detector.

9.3.1 Routine measurement of ISL

The cut-off filter is satisfactory for the majority of routine applications. It must always be borne in mind that the ISL is a function of the sample: the measurement of x% ISL with a cut-off filter does not mean that x% will again be present when a different absorber is in the beam. It is better to regard the filter method as one which detects stray-light rather than measures it.

The solutions and liquids listed below are recommended as standard 'cut-off' filters. They are also the recommendations of the American Society for Testing and Materials (ASTM), and are generally accepted as industrial standards [11]. Compared with glass filters, they have the advantages of reproducibility and freedom from fluorescence. It should be noted that the cells used must be clean, free from fluorescence and with as high a transmission as possible in the region under investigation. Attention to these factors is particularly important when measuring stray-light below 220 nm.

The absorbance of the first two of these filters is affected by dissolved oxygen. Pure nitrogen should be bubbled through for several minutes before use and the water should be freshly distilled. Water purified by ion-exchange methods may contain significant amounts of organic impurities. Different concentrations or pathlengths may be used to displace the absorption edge of these filters so that they can be used in other regions. The absorbances of the solutions increase with temperature by about 2% per °C [12].

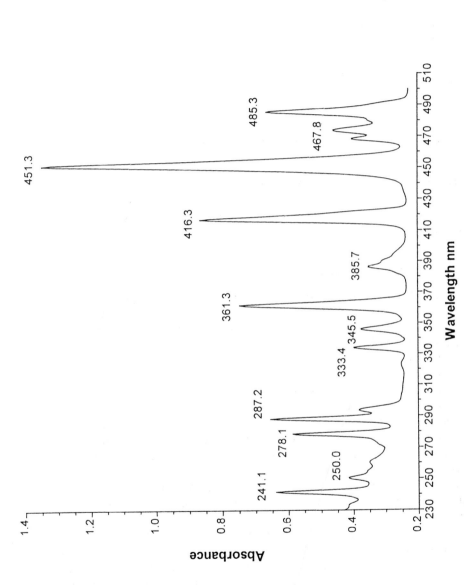

Fig. 9.7 Holmium oxide (5%w/v) in 10%v/v perchloric acid solution, 10 mm pathlength and 1 nm SBW. Data values are from NIST [5]. Jasco V560 spectrometer, scan rate 40 nm min^{-1}, medium detector response and data pitch 0.1 nm.

138 *Standards in Absorption Spectrometry*

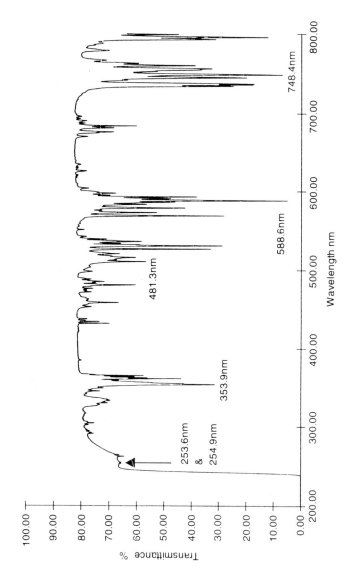

Fig. 9.8 McCrone filter, 200–800 nm, Cary 5, 1 nm SBW [9].

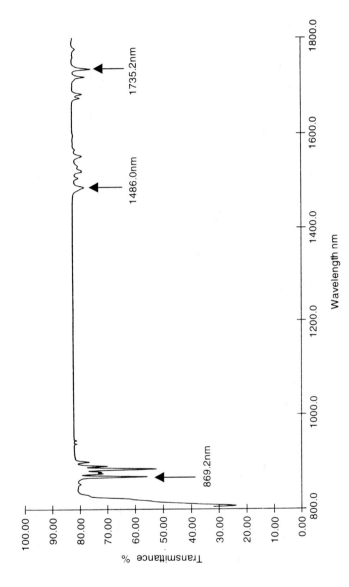

Fig. 9.9 McCrone filter, 800–1800 nm, Cary 5, 1 nm SBW [10].

Table 9.2 Cut-off filters for stray-light tests. Pathlength 10 mm.

Spectral range (nm)	Liquid or solution
170–183.5	Water
175–200	Aqueous KCl (12 g l^{-1})
195–223	Aqueous NaBr (10 g l^{-1})
210–259	Aqueous NaI (10 g l^{-1})
250–320	Acetone
300–385	Aqueous NaNO$_2$ (50 g l^{-1})

NIST supplies crystalline potassium iodide (SRM 2032) as a standard for the assessment of stray-light. Absorptivity values are given at 5 nm intervals from 240 to 275 nm.

When measuring very low stray-light levels, it is necessary to attenuate the reference beam with a metal screen in order to extend the absorbance range of the instrument. Details of this procedure are given by the ASTM [11]. For the measurement of stray-light above 700 nm, a filter is required whose cut-off region extends to longer wavelengths. There are currently no suitable liquid or solution filters, so a glass filter must be employed. Heat-absorbing glass is fairly satisfactory; a sample tested had a rather broad cut-off extending from 82% transmittance at 600 nm to less than 0.5% at 1000 nm. This is far from ideal; nevertheless the filter gave a useful indication of stray-light at 900 nm in two single-monochromator grating spectrophotometers, which proved to be 15–20% in each case. To evaluate a filter for use above 700 nm, its true absorbance must be measured by a spectrometer equipped with a lead sulphide detector.

9.3.2 Quick stray-light checks

Vycor glass has been used extensively as a cut-off filter for stray-light measurement in the range 200–210 nm, but it is unsuitable as a standard because its absorption edge is not very steep, different samples may not have exactly the same absorption characteristics and there may be interference from the fluorescence of the glass. However, it is useful as a check in this spectral range when the stray-light is relatively high, i.e. greater than 1%.

9.4 Absorbance standards

9.4.1 Photometric linearity

Several linearity standards were mentioned in Chapter 4, e.g. potassium dichromate for the UV region or green food-dye for the visible region. A single concentration can be used in a series of cells of increasing pathlength, e.g. 5–40 mm. This procedure is recommended since the tolerance on cell pathlength is far smaller than the uncertainties associated with volumetric manipulations. Additionally, it is not necessary to assume Beer's law. It is important to note that photometric linearity measured in this way does not always ensure detector linearity. For a fuller discussion of this topic, see Clarke [13].

9.4.2 Photometric accuracy

High-accuracy calibration

For the most exacting work, a set of neutral glass filters or metal-on-fused-quartz filters calibrated by the NIST or NPL is recommended. The conditions for their use will be detailed by the calibration service. However, problems with inter-reflection errors and cleaning of the metal-on-fused-quartz filters means that they have to be used and handled only with the greatest care.

Routine calibration

It is recommended that potassium dichromate in 0.005 M sulphuric acid is used as the standard. Potassium dichromate is readily available with a certified 'as is' purity of >99.9%. The water content is so low that drying, recommended by some authors, is probably unnecessary. Alternatively, potassium dichromate can be purchased from NIST as SRM 935 (see Section 4.5.5). Measurements should be made in 10 mm cells with the temperature controlled in the range 15–25°C, using 0.005 M sulphuric acid as the reference. Table 9.3 gives the expected values for the two solutions at the two maxima and minima of the solutions based on literature values. The tolerances represent the range of acceptable values, based on the uncertainties of the literature values and the temperature coefficient over this temperature range.

Table 9.3 Recommended absorbance values for the acidic potassium dichromate solutions. See Fig. 4.1 for spectrum.

| Wavelength | UVSG recommended values [15] | | European Pharmacopoeia [16] | |
	A (1%, 1 cm)	Maximum tolerance	A (1%, 1 cm)	Maximum tolerance
235	125.1	123.2–127.0	124.5	122.9–126.2
257	145.4	143.9–146.9	144.0	142.4–145.7
313	48.8	48.1–49.5	48.6	47.0–50.3
350	107.1	106.0–108.2	106.6	104.9–108.2

UVSG recommended values are for a concentration range of 50–100 mg per litre of potassium dichromate in 0.005 M sulphuric acid and measured at a temperature of $20\pm5°C$. The European Pharmacopoeia values are for a solution of 57–63 mg per litre of potassium dichromate in 0.005 M sulphuric acid and measured at an unspecified temperature. However, in the general requirements for the measurement of absorbance, a temperature of $20°\pm1C$ is specified [14]. The EP values considerably overlap our recommendations, but the mean values are slightly low.

References

1. Verrill, J.F. (1987) In: *Advances in Standards and Methodology in Spectrophotometry*, (eds C. Burgess and K.D. Mielenz), p. 113. Elsevier Science.
2. PharmEuropa, Special Issue. (1996) *Technical Guide for the Evaluation of Monographs*, 2nd ed. p. 12.
3. Burgess, C. Unpublished work.
4. Lee Smith, A. (1981) In: *Treatise on Analytical Chemistry*, (eds I.M. Kothoff and P.J. Elving), 2nd edn. Part 1, Vol. 7, p. 283, 284. Wiley.
5. NBS Special Publication 260–102 (1986).
6. Lang, L. (ed.) (1966) *Absorption Spectra in the UV and Visible Region*, Vol. 1, 4th ed, pp. 367–369. Akademiai Kiado, Budapest.
7. D.R. Lide (ed.) (1996–1997) Line spectra of the elements. In: *Chemical Rubber Handbook*, 77th ed. pp. 10–57.
8. ATI-Unicam Ltd (1994) Adapted from figure 14. *Regulatory Compliance; A Guide to UV-Visible Spectrometer System Validation*. ATI-Unicam Limited.
9. Adapted from NPL reference spectra, JN96.
10. Adapted from NPL reference spectra, JN96.
11. ASTM Standard E-387-84 (1997) (Reapproved 1995). *Annual Book of ASTM Standards*, Section 03.06.
12. Slavin, W. (1963) *Anal Chem.*, **35**, 561.
13. Clarke, F.J.J. (1981) *UV Group Bulletin*, **9**(2), 81–90.
14. British Pharmacopoeia (1998) Appendix IIB, A117.
15. Values from 1st ed as confirmed independently by Escolar *et al.* (1986) *Applied Spectroscopy*, **40**(8), 1160.
16. European Pharmacopoeia, 3rd ed. (1997).

Part 2
Practical Absorption Spectrometry

10 Absorption Spectrometry

10.1 Absorption spectrometry in the ultraviolet and visible regions

It is now 50 years since the first spectrometers working in the ultraviolet and visible (UV–VIS) regions of the spectrum came into general use, and over this period they have become the most important analytical instrument in many chemical, biological and clinical laboratories. Because the technique has become so commonplace, it is assumed that every scientist knows how to 'run' an absorption spectrum. However, proper training in the technique is essential for there are many pitfalls to be avoided if reliable results are to be produced. Although this book is intended to be an introduction to those new to the technique, it may also help to improve the results of more experienced users.

The potential of absorption spectrometry is best described by outlining some of its merits and limitations.

10.1.1 Characterization of compounds

Most organic compounds and many inorganic ions and complexes absorb radiation in the UV–VIS region. A plot of this absorption by a compound against wavelength is called its absorption spectrum; this has a shape that is characteristic of a particular compound or class of compounds. The UV–VIS spectrum does not usually give enough information to identify an unknown compound, but when combined with other analytical techniques or with chromatographic separation, an unequivocal identification can be achieved, Absorption spectrometry is a non-destructive technique and is extremely sensitive, and is therefore ideal for the characterization of small amounts of precious compounds. As an extreme example, it is used for the measurement of the pigments in the single retinal receptor cells. A custom-built microspectrometer is used which is capable of recording satisfactory spectra from only 4×10^{-16} mol pigment, which is only about 10^8 molecules.

10.1.2 Quantitative assay

The most important application of the technique is as a means of measuring concentration, and modern instruments are designed to facilitate rapid and accurate measurements. The precision that can be achieved depends upon a number of factors that will be discussed later in this volume, but single measurements of precision better than $\pm\,0.5\%$ should be possible. One particularly important merit of the technique is that trace components can be measured in the presence of high concentrations of other components if there is a sufficient difference in their absorption spectra. Similarly, mixtures of compounds with differing spectra can be analysed, and methods for doing this will be discussed in Chapter 11.

The technique is best suited to dilute solutions, though gases and solids can be measured by special methods. Thus, solubility in a suitable solvent is a prerequisite for the accurate measurement of a particular sample. The optimum concentration range for the measurement of a compound is limited and does not generally exceed a 100-fold range, and so it may be necessary to adjust the concentration of the test solution to bring it into the optimum range. Beyond manipulations of the concentration of solutions, measuring the UV–VIS absorption will not change the sample in any way, and even UV-sensitive compounds will suffer negligible damage, for the amount of radiation absorbed by the sample during the measurement is very small.

10.1.3 Rapid assays

The speed of photodetectors and modern electronic components means that the measurement of solutes in flowing or rapidly changing systems can be carried out with millisecond time resolution. A popular application is the monitoring of the eluent from HPLC columns; the concentration of a particular component can be followed by measuring the absorbance at a suitable wavelength, or compounds eluted from the column can be identified by making rapid spectral scans.

Absorption measurements are also the most popular means of following the kinetics of reaction systems, since they do not interfere with the progress of the reaction in any way. The 'stopped flow' technique is an important method for the study of biological reactions, and light absorption is the most popular monitoring technique since it has both speed and selectivity for a particular component of a reaction mixture. Probably the most advanced kinetic monitoring systems are found in flash photolysis systems where reactions with lifetimes of the order of 10^{-8} s can be followed. By using very fast detection systems, e.g. streak cameras, absorption measurements in the picosecond time domain are possible. Another important application of rapid absorption measurement is in the clinical field, where colorimetric assays have been worked out for many biologically important compounds. Continuous flow systems employing these reactions can measure many hundreds of samples in an hour.

All of these techniques, however advanced, are based on the same fundamental principles of spectrometry, and a sound grasp of these particular principles is essential, whatever kind of measurement you intend to make. The remainder of this chapter will introduce the fundamentals of absorption spectrometry, and the rest of the book will be concerned with the practicalities of their application.

10.2 The ultraviolet and visible spectrum

Figure 10.1 shows the electromagnetic spectrum and the narrow region that is of interest to us. Despite its narrowness, this band of radiation is vital to life on earth because its interaction with molecules is the primary step in both photosynthesis and vision. Measurement of such molecular interactions forms the basis of UV–VIS spectroscopy and can provide a wealth of information about the molecules. The diagram shows that radiation can be measured in terms of either wavelength or frequency. In the UV–VIS region, wavelength is generally used, even though frequency would be more appropriate. The convenient unit of wavelength in this region is the nanometre (nm, 10^{-9}m, formerly called the millimicron, mµ). When frequency is used, it is generally expressed in terms of wavenumbers – the number of waves per cm – rather than the number of waves per unit time, thus avoiding any assumptions about the speed of light in a particular medium. These quantities are related by:

$$\lambda = \frac{10^9 c}{\nu} = \frac{10^7}{\bar{\nu}}$$

where λ is the wavelength in nm, ν is the frequency in Hz, $\bar{\nu}$ is the frequency in wavenumbers (cm^{-1}) and c m s^{-1} is the velocity of light (2.998×10^8 m s^{-1} in air).

The limits of the visible spectrum are ill-defined, but most instrument makers take it to lie between 400 and 800 nm; only radiation lying between these limits can properly be described as 'light'. The ultraviolet region extends from 400 nm down to 100 nm. Since most gases have appreciable absorption below 185 nm, measurements in the range below this wavelength can only be made using instruments in evacuated enclosures; this vacuum–UV region is therefore outside the range of most instruments and will not be dealt with in this book. The remainder of the UV region is loosely divided into the near-UV and far-UV regions, though there is no consensus on where the dividing line should be, and different authorities place it at various points between 200 and 300 nm. These terms will only be used in a relative sense in this text. The near infrared is usually defined as 800–2500 nm.

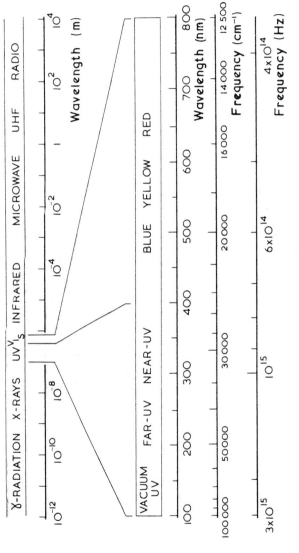

Fig. 10.1 The range of electromagnetic radiation. The upper part is plotted on a logarithmic scale and illustrates the small extent of the UV–VIS region. The lower part is an enlargement of this region on a linear wavelength scale and shows the relationship of this scale to the frequency scales.

10.3 The absorption of radiation

When UV or VIS radiation encounters an atom or molecule, an interaction between the radiation and the electrons of the latter may take place. This absorption process is very specific and results in an attenuation of the radiation and an increase in the energy of the electrons of the atom or molecule. This may be regarded as the promotion of one of the outer or bonding electrons from a 'ground-state' energy level into one of higher energy (Fig. 10.2). These levels are separated by a discrete energy increment, E, which is determined by the nature of the atom or molecule, and only parcels of radiation of energy E can be absorbed. This parcel of radiation is termed a quantum and its energy is related to the frequency and wavelength of the radiation by:

$$E = h\nu = \frac{hc}{\lambda} \times 10^9$$

where h is Planck's constant (6.63×10^{-34} J s), c is the velocity of light (2.998×10^8 m s^{-1}) and λ is in nm.

Suppose that Fig. 10.2 represents the energy level diagram of a molecule and that $E = 7.95 \times 10^{-19}$ J which is equivalent to $\lambda = 250$ nm. If the molecule is exposed to a complete spectrum of UV and VIS radiation, only that of wavelength exactly 250 nm will be absorbed. A plot of absorption versus wavelength – the absorption spectrum – would be a single sharp line, as shown in Fig. 10.3. In reality, even the simplest molecules have large numbers of energy levels and their absorption spectra are far more complex than this. In addition, each electronic energy level has a group of closely spaced vibrational levels associated with it due to small increments of the energy of the molecule caused by the relative motions of its constituent atoms. These vibrational levels

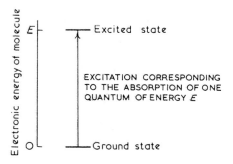

Fig. 10.2 A representation of the absorption of radiation by a molecule. This results in the excitation of one electron from the ground-state to an excited state of energy E.

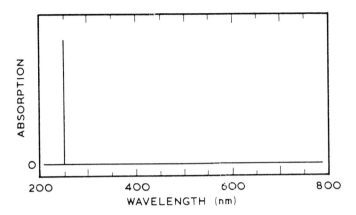

Fig. 10.3 The hypothetical absorption spectrum for a molecule having the single absorption process shown in Fig. 10.2.

overlap to such an extent that most spectrometers are incapable of resolving them and the measured spectrum appears as a broad bell-shaped peak.

Benzene is a molecule comprised of 12 atoms and its absorption spectrum is relatively complex. Figure 10.4(a) shows a part of the UV spectrum and has a group of minor peaks that form an absorption band representing a major type of electronic excitation. The minor peaks represent small differences in the energies of electronic excitations within the group, while the rounded shape of each of these minor bands is due to the fusion of the myriad vibrational levels associated with each electronic transition. The composition of these minor bands becomes more apparent if the benzene spectrum is measured with the sample in gaseous form rather than in solution. Figure 10.5 shows the 240–265 nm region of the vapour-phase spectrum run on a conventional spectrometer under conditions of maximum resolution. This shows how each of the four maxima seen in this region in the solution spectrum can be resolved into a major peak followed by a series of lesser ones of decreasing energy. Even this vapour-phase spectrum cannot be regarded as an 'absolute' spectrum of benzene, for an ultra-high-resolution spectrometer would be capable of further separating these peaks into even narrower ones.

Although appearing very complex, the benzene spectrum is relatively simple because of the great symmetry of the molecule. The majority of organic compounds have so many overlapping bands that these merge into one or two broad maxima when their spectra are measured in solution. The spectrum of aniline [Fig. 10.6(a)] is thus more typical of organic compounds in general and also serves to illustrate how simple substitution of the benzene molecule can transform its spectrum.

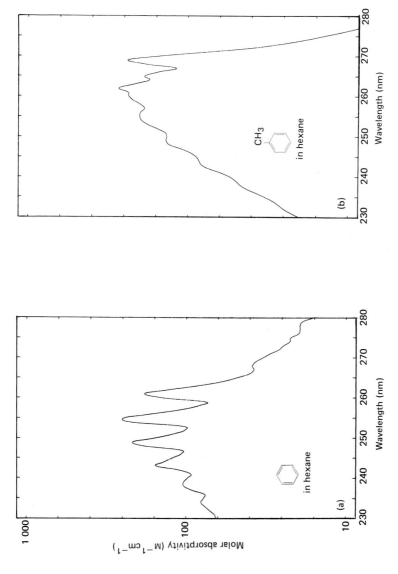

Fig. 10.4 UV absorption spectra of (a) benzene and (b) toluene dissolved in hexane.

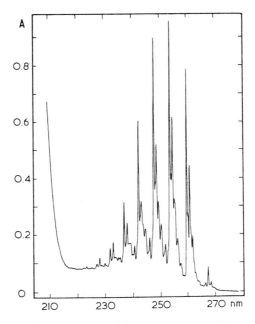

Fig. 10.5 An absorption spectrum of benzene vapour. Air was saturated with vapour at 1 atm and measured in 5 mm pathlength. ESW is about 0.03 nm.

10.4 Molecular structure and absorption spectra

This is a vast subject that can only be briefly outlined here. Despite the great amount of effort that has been put into the prediction of absorption spectra by the calculation of molecular energy levels, most discussions of structure – spectra correlations are based on empirical rules, and these are always hedged around by provisos and qualifications. Most molecules absorb somewhere in the UV–VIS region and, in general, the more complex the molecule, the longer the wavelength of its first absorption band, i.e. the band of lowest energy and longest wavelength. Thus, the simplest molecules, e.g. O_2, absorb appreciably only below 190 nm, while the first band of a complex molecule like methylene blue lies at the red end of the visible region.

The relationship of the absorption spectra of organic compounds to their structure has been extensively studied and is, to some extent, understood. The first step in predicting the absorption spectrum of a molecule is to consider its bonding electrons. The outer electrons of organic compounds are of three main types: σ-electrons which are involved in covalent bonds, π-electrons which are involved in double and triple bonds, and *n*-electrons which are the non-bonding electrons associated with hetero atoms, e.g. nitrogen and oxygen. Current

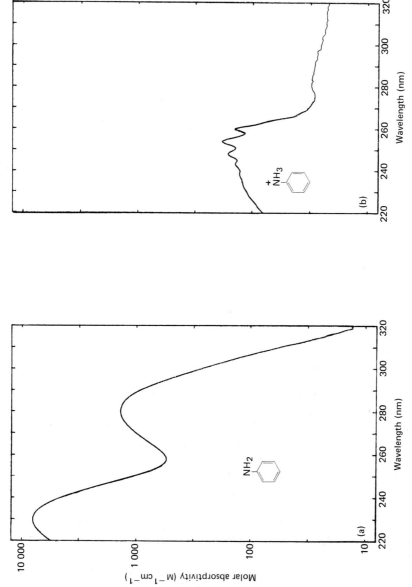

Fig. 10.6 UV absorption spectra of (a) aniline and (b) the anilinium ion in aqueous solution.

explanations of absorption spectra are based on the assignment of the spectral bands to the excitations of these different classes of electron.

Excitation of σ-electrons requires the highest energies. Since all organic compounds have σ-electrons, they all absorb UV radiation, but small saturated molecules, e.g. the smaller alkanes, only show appreciable absorption below 200 nm. Transitions involving π-electrons are found at longer wavelengths, thus an unsaturated compound will always have a first absorption band at a longer wavelength than a similar saturated compound. This is exemplified by Fig. 10.4(a); hexane has no π-electrons and so absorbs at shorter wavelengths than benzene and could thus be used as a non-absorbing solvent for the latter. Not only the size of the molecule, but the number of π-electrons forming the conjugated system has a profound effect upon the location (λ_{max}) of the first band. This is illustrated in Fig. 10.7, where increasing the number of conjugated rings is seen to displace the first band from the UV into the visible region, naphthacene appearing coloured.

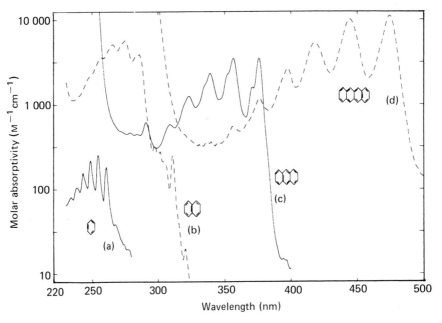

Fig. 10.7 Absorption spectra of (a) benzene, (b) naphthalene, (c) anthracene in hexane and (d) naphthacene in toluene.

10.4.1 *Substituent effects*

The absorption spectra of compounds formed from a hydrocarbon by the simple substitution of hydrogen atoms by other groups can to some extent be predicted, since a given substituent has a known effect upon the electrons of the molecule and consequently its absorption spectrum. For example, alkyl groups have only minor effects upon the electronic orbitals of a molecule and hence little effect upon the absorption bands. Figure 10.4(b) shows the absorption spectrum of toluene: comparison with Fig. 10.4(a) shows that the absorption bands of benzene have been smoothed by the introduction of further minor energy levels, and the centres of the bands have been shifted to longer wavelength. This displacement is described as a bathochromic shift. Substituents containing hetero atoms generally have greater effects than alkyl substituents, as illustrated by the spectrum of aniline [Fig. 10.6(a)]. In addition, if these atoms can take up or lose an electron, further major changes in the absorption spectrum are seen. Thus, if an aniline solution is acidified, a proton is taken up to form the anilinium ion [Fig. 10.6(b)]. The effect of tying up the nitrogen lone pair electrons in this way is to shift the bands to shorter wavelengths (an hypsochromic shift) and the spectrum becomes very similar to that of toluene.

Substituents containing $C=C$, $C=N$ or $C=O$ double bonds in positions where they can conjugate with double bonds in the core molecule will cause much larger hypsochromic shifts than alkyl substituents, and may also introduce extra characteristic bands of their own. Thus, many carbonyl compounds have a band at about 280 nm due to the excitation of an oxygen lone pair electron into an unoccupied π-orbital: it is therefore designated an $n \rightarrow \pi^*$ transition. This band is the main feature in the spectrum of butanone shown in Fig. 10.8. The dominance of each characteristic band in the spectra of small molecules can be of help in classifying an unknown compound, but it is not generally possible to distinguish between compounds in a given class. For example, the spectra of the naturally occurring pyrimidine bases are very similar and cannot be used to make positive identification. However, in this case, individual compounds can be characterized by means of their pK, and this can be conveniently determined by UV absorption measurements. Ionization of the compounds results in a major spectral change (Fig. 10.9) and so measurements of the spectra over a range of pH values will enable the pK to be accurately evaluated.

10.4.2 Solvent effects

A change in the pH of the medium generally has a profound effect upon the spectrum of a molecule containing hetero atoms. Beyond this effect, a change in the polarity of the solvent can cause smaller shifts in the spectra of most compounds. An increase in solvent polarity will cause a bathochromic shift of $\pi \rightarrow \pi^*$ absorption bands, though not all of the band systems in a particular molecule may be affected to the same extent, and so the shape of the spectrum may well change. $n \rightarrow \pi^*$ transitions are also sensitive to solvent polarity, but undergo hypochromic shifts when the

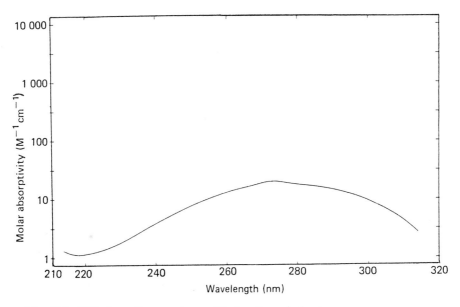

Fig. 10.8 UV absorption spectrum of butan-2-one in heptane.

polarity is increased. Table 10.1 illustrates this effect on the 280 nm absorption band of a typical carbonyl compound; other solvent properties beyond polarity (measured as dielectric constant) are involved in the observed shifts, e.g. the ability of the solvent molecules to form hydrogen bonds, but the relationship is evident. Since the effect of changing solvent differs for $\pi \to \pi^*$ and $n \to \pi^*$ bands, compounds having both types of bands will show major changes in their spectral shapes. The possibility of solvent effects must be remembered when comparing spectra measured in the two common spectroscopic solvents, hexane and ethanol, which differ greatly in polarity. As Table 10.1 shows, the absorption maximum of acetone is shifted by 9 nm on going from one to the other.

10.5 Quantitative absorption spectrometry

Beyond having characteristic peak positions, the absorption spectrum of a given compound also has characteristic peak heights which can serve as an additional aid to identification and, more important, will form the basis of a quantitative assay. When radiation travels through the solution of an absorbing compound it is reduced in intensity by each molecule that it encounters according to an exponential law. The amount of radiation absorbed by a solution is thus an exponential function of the concentration of the solution and the distance that the radiation passes through it (Fig. 10.10). In practice, it is the amount of radiation transmitted by the solution

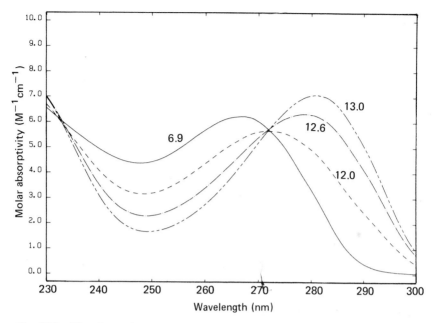

Fig. 10.9 The absorption spectrum of an aqueous solution of cytosine at the pH values shown. The spectra at pH = 6.9 and 13.0 are taken to represent the protonated and unprotonated molecules, respectively, and the spectra at intermediate pH represent mixtures of the two species. Both species have the same molar absorptivity at about 234 and 272 nm; consequently, the spectra all cross at these wavelengths forming isosbestic points. From a series of such curves, the pK_a can be shown to be 12.2.

Table 10.1 The effect of solvent polarity upon the $n \rightarrow \pi^*$ absorption band of acetone. Data from Ref. [1].

Solvent	Dielectric constant	λ_{max} (nm)	Hypsochromic shift $\Delta \bar{\nu}$ (cm^{-1})
Hexane	1.9	280	0
Carbon tetrachloride	2.2	280	0
Dioxane	2.2	277	300
Chloroform	4.8	276	400
Dimethylformamide	38.4	275	500
Dimethyl sulphoxide	44.6	274	600
Ethanol	24.3	271	1000
Methanol	32.6	270	1200
Acetic acid	6.2	267	1600
Water	78.5	265	2000

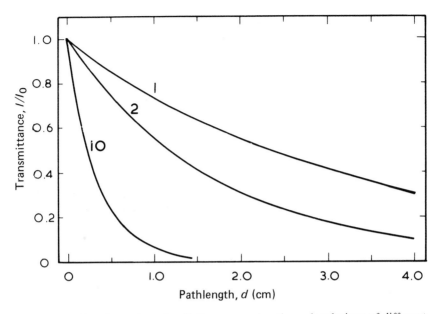

Fig. 10.10 The absorption of radiation on passing through solutions of different thicknesses and concentrations. The curves show the attenuation of the measuring beam when passing through solutions of an absorbing solute at relative concentrations of 1, 2 and 10.

that is measured. This is expressed as the transmittance, which is the ratio of the amount of radiation transmitted to that incident upon the solution:

$$T = \frac{I}{I_0} \propto e^{-bc}$$

where T is the transmittance of the solution, I_0 is the intensity of radiation entering and I the intensity leaving a solution of optical pathlength b cm and concentration c mol l^{-1}. Converting to logarithms and changing the signs:

$$-\log_{10} T = \log_{10} \frac{I_0}{I} \propto bc$$

$\log_{10} I_0/I$ is termed the absorbance or optical density of the solution, and the relationship is called the Beer–Lambert law or Beer's law:

$$A = -\log_{10} T = \epsilon bc$$

where A is the absorbance of a particular wavelength and is dimensionless, and the constant of proportionality (ϵ) is termed the *molar absorptivity* at the specified wavelength and has units of $M^{-1}\,cm^{-1}$ (M stands for molarity or the molecular weight in g l^{-1}). The molar absorptivity thus represents the absorbance of a 1 M solution measured in a 10 mm layer. Occasionally, reference is made to the 'extinction' values of a solution expressed as $A_{1\,cm}^{1\%}$ or $E_{1\,cm}^{1\%}$, which means the absorbance of a 1% w/v solution of the solute in a 1 cm pathlength. For a compound of molecular weight m, a 1% solution is $10/m$ M, and so:

$$A_{1\,cm}^{1\%} = \frac{10 \times \epsilon}{m}$$

On the other hand, a few recent texts used SI units where concentration is expressed in mol m^{-3} and the molar absorptivity is for a pathlength of 1 m. The SI molar absorptivity is related to ϵ by:

$$\varepsilon' = \geqslant.\varnothing\epsilon \text{ mol}^{-\varnothing} \text{ m}$$

Examination of the spectra given earlier in this chapter shows that the ϵ values of maxima generally fall in the range 10–10^5 $M^{-1}\,cm^{-1}$. An absorbance of 0.3 A can be measured accurately on most instruments: a compound of $\epsilon_{max} = 10^4$ $M^{-1}cm^{-1}$ in solution at a concentration of 30 µM measured in a 1 cm pathlength cell will have this absorbance. One millilitre of solution should suffice for this measurement, and so 30 nmol of the compound can be readily measured. By using a microcell and optimizing the conditions of measurement, the detection limit for such a compound can probably be extended into the pmol region.

10.6 Measurement of absorption spectra

The task is to measure the fraction of radiation entering the cell that is absorbed by the sample, to convert these data into absorbance values and, in the case of recording instruments, to present these values in the form of a continuous spectrum. Most instruments employ a phototube or photoconductive detector whose response is linearly related to light intensity. The instrument is arranged to measure I, the radiation intensity transmitted by the sample, and compares it to I_0, the intensity of a reference beam that does not pass through the sample. This comparison gives the transmittance, $T = I/I_0$. Conversion of T to an absorbance value was a major problem in earlier instruments, since it involves a logarithmic function, i.e. $A = -\log T$, and ingenious mechanical devices or logarithmic recorder slidewires were employed. The advent of cheap solid-state operational amplifiers for linear–log conversion, and more recently digital devices, means that the conversion is performed very accurately in modern instruments.

The fundamental problems involved in measuring the transmission are two-fold: (a) to ensure that only radiation of the specified wavelength reaches the detector; and (b) to ensure that the attenuation of the radiation is due solely to absorption by the solute. The problem associated with (a) will be discussed later in the chapter; as far as (b) is concerned, the factors causing attentuation of the measuring beam are illustrated in Fig. 10.11 and must be compensated for when measurements are made.

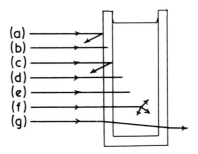

Fig. 10.11 Possible causes of attenuation of the measuring beam as it passes through a cell containing a sample solution. (a) Reflection at air/glass interface; (b) absorption by glass; (c) reflection at glass/liquid interface; (d) absorption by solute; (e) absorption by solvent; (f) scatter by solution; (g) refraction or dispersion by cell. Accurate measurements of the solute absorbance (d) require the other factors to be minimized or compensated for.

The earliest absorption spectra were recorded by passing 'white' radiation through the sample and into a spectrograph which dispersed the radiation by means of a prism and recorded the resulting spectrum on a photographic plate (Fig. 10.12). Beyond the problems of developing the plate, the principal difficulties were in compensating for fluctuations in the source, correcting for absorption, scatter and reflection by the cell and solvent, and calibrating the response of the plate to radiation of different wavelengths and intensities. The development of the phototube, which has a linear relationship between its response and the light intensity, was the key to the development of new instruments for the

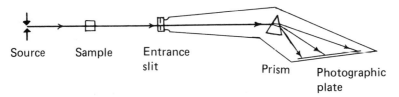

Fig. 10.12 Schematic diagram of the arrangement for measuring absorption spectra by means of a spectrograph.

measurement of absorption spectra. The first commercial instruments appeared in the 1940s, and since they were concerned with the measurement of both wavelength and intensity, were termed 'spectrophotometers'. Modern usage has contracted this to spectrometer, and the technique has become spectrometry. The design and construction of spectrometers has been dealt with in Chapter 3, but two properties of spectrometers that are absolutely vital to the proper measurement of absorption spectra, i.e. spectral bandwidth and stray-light, will be introduced here.

10.6.1 Bandwidth

The width of the entrance and exit slits of a monochromator determines the spread of wavelengths that emerge from it – the bandwidth of the radiation. In addition, the amount of energy passing through the instrument is also determined by the slitwidth. When measuring an absorption spectrum, the bandwidth should be as small as possible to attain the maximum spectral resolution; ideally when the monochromator is set to, say 260 nm, only radiation in a narrow spectral range centred at 260 nm should emerge from the exit slit. This is illustrated in Fig. 10.13, where (a) shows the ideal narrow rectangular profile. In reality, the

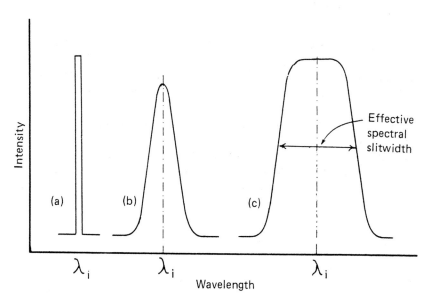

Fig. 10.13 Spectral distribution of radiation emerging from a monochromator set at indicated wavelength λ_i. (a) Ideal profile; (b) typical profile for small slit settings; (c) profile for practical slit settings.

spectral distribution from a monochromator with narrow slits is roughly triangular in shape, as in (b). However, even this profile cannot be used in practice for a slit opening given this profile is unlikely to pass sufficient energy for the efficient operation of the instrument, and so the slits must be opened further, and a typical operating profile is shown in (c). The bandwidth is best expressed as the width of the profile at half-peak height and is termed the effective spectral slitwidth (ESW). If the ESW approaches the width of the absorption peak to be measured, then serious degradation of the recorded spectrum will result. The true width of an absorption band at half-peak height is termed the natural bandwidth (NBW), and Fig. 10.14 shows how the measured height of an absorption band falls as the ratio ESW/NBW is increased. As a rule-of-thumb, the error in peak height measurement will be negligible if the ESW is kept at less than 10% of the NBW of the band.

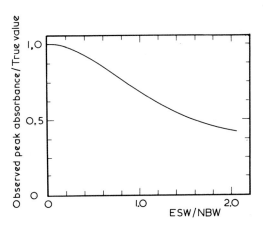

Fig. 10.14 Effect of slit setting upon the measured height of a Gaussian-shaped absorption band.

When compounds with closely spaced absorption maxima are measured, the peak-broadening effect due to excessive ESW will cause the individual peaks to merge into a broad band. This partially explains the differences between the benzene spectra of Figs 10.4 and 10.5. The loss of some of the peaks seen in the gas phase spectrum when the molecule was put into hexane is due in part to solvent broadening of the lines, but is also due to the greater ESW used in measuring the solution spectrum. Quantitative measurements of spectra with such narrow bands should be avoided, but if they must be made, a compromise has to be reached in selecting an ESW that is sufficiently small to resolve the bands satisfactorily while allowing sufficient energy to pass through the sample to ensure photometric accuracy.

10.6.2 Stray-light

When 'white' radiation from a tungsten or deuterium lamp is passed through an ideal monochromator, only the selected wavelength will emerge from the exit slit. Figure 10.13 showed that in normal operation, a band of wavelengths a few nanometres wide is found, but in addition to this, optical defects in real monochromators cause radiation of widely differing wavelengths to emerge. Such emergent radiation of wavelengths other than the indicated wavelength is termed stray-light, even though it may be in the UV region. It is an inherent property of diffraction gratings that they produce several overlapping spectra of different orders, and so when set at an angle to reflect 750 nm light, the grating will also reflect radiation of 375 and 250 nm. Consequently, this effect is a major cause of stray-light in grating monochromators. Beyond this, imperfections in gratings, prisms or mirrors, reflection or refraction at slits and beam masks, and dust contamination will all contribute to the stray-light.

To demonstrate these effects, turn the wavelength setting of your spectrometer to 550 nm and open the slits as wide as possible. Place a piece of white non-fluorescent card in the sample position and note how the colour of the patch of light on the card changes from side-to-side; this is due primarily to the width of the slit and is termed near stray-light. Turn the wavelength down to 380 nm: the patch of light should disappear before this wavelength is reached, but you will probably still be able to see a patch of faint white far stray-light. This is of far lower intensity than the near stray-light but can cause far more serious distortions of spectra. Suppose that when the instrument is set at 380 nm, 1% of the radiation reaching the detector is stray-light of wavelengths greater than 400 nm. If a sample is placed in the beam which transmits only 10% of the 380 nm radiation but has effectively 100% transmission above 400 nm, then the stray-light will form 9% of the light reaching the detector. If the true absorbance of the sample at 380 nm is 1.000 A, the apparent absorbance will only be 0.963 A.

This effect becomes more significant when making measurements at shorter wavelengths, for the intensity of the radiation from the lamp is less at short wavelengths and so the stray-light will be a greater proportion of the radiation reaching the detector. In addition, the sensitivity of the detector is less at shorter wavelengths and the effect of long-wave stray-light will be proportionately greater. The possibility of significant errors caused by stray-light should be borne in mind when using an instrument with a tungsten lamp below 400 nm and with a deuterium lamp below 250 nm. Prism instruments are generally fitted with a UV-transmitting filter that is inserted into the beam below 400 nm, but there is no suitable filter transmitting only below 250 nm for use with a deuterium source. Grating instruments must have accessory filters to remove spectra of unwanted orders, but again restrictions on the choice of filter means that stray-light becomes a problem at shorter wavelengths.

The effects of stray-light are considered in detail in Chapter 6.

Reference

1. Perkampus, H.H. (1992) *UV Atlas of Organic Compounds*, 2nd edn (2 volumes). VCH.

General reading

Stern, E.S. and Timmons, C.J. (1970) *Introduction to Electronic Absorption Spectroscopy in Organic Chemistry*, 3rd edn. Arnold, London.
Steward, J.E. (ed.) (1975) *Introduction to Ultraviolet and Visible Spectroscopy*. Pye Unicam, Cambridge.
Beaven, G.H., Johnson, E.A., Willis, H.A. and Miller, R.G.J. (1961) *Molecular Spectroscopy*. Heywood, London.
Anon. (1973) *Optimum Parameters for Spectrophotometry*. Varian Instruments, Palo Alto, California.
Edisbury, J.R. (1966) *Practical Hints on Absorption Spectrometry*. Hilger, London.
Rao, C.N.R. (1967) *Ultraviolet and Visible Spectroscopy*, 2nd edn. Butterworths, London.

11 Measuring the Spectrum

This chapter is primarily concerned with the measurement of the absorption spectra of clear solutions; techniques for dealing with more difficult samples will be briefly dealt with in Section 11.8.

11.1 Choice of solvent

When presented with a solid sample, a determined attempt should be made to get all of it into solution. The nature of the compound should be considered. For organic compounds, the rule of 'like dissolves like' should be remembered, and so the polarity of the solvent should match that of the sample. Thus, a hydrocarbon will probably only dissolve in a hydrocarbon solvent, while polar molecules require polar solvents. A solvent can be found for most low-molecular weight organic molecules. The properties of some useful solvents are given in Ref. [13]. Higher molecular weight compounds present greater problems, and some proteins and polymers will not form true solutions in any solvent. Many biological macromolecules will give clear micellar suspensions in detergents. If you are unsure about the choice of a suitable solvent, do not hesitate to refer to the originator of the sample – he or she is the person most likely to understand the problem.

Beyond the fundamental question of its ability to dissolve the samples, the following points should also be considered when choosing a solvent.

11.1.1 Transmission

The solvent must have a high transmittance in the spectral region of interest. This is particularly important when selecting polar organic solvents for use below 300 nm. The useful ranges of some common solvents are given in Ref. [13]. Although distilled water transmits satisfactorily down below 200 nm, it should be remembered that detergents and buffer salts can absorb strongly at wavelengths up to 300 nm. Other less obvious transmission problems should also be borne in mind. Many organic solvents contain antioxidants or inhibitors that absorb at longer wavelengths than the solvent itself, and so, if the latter is allowed to evaporate, the apparent cut-off wavelength can change. Oxygen and other gases dissolve in water and many organic solvents, and this will affect solvent transmission at shorter wavelengths; consequently, routine purging of solutions with nitrogen or argon is advisable for measurements in the far-UV.

11.1.2 Handling problems

Highly volatile solvents, e.g. diethyl ether, should be avoided at all costs, and alkanes smaller than hexane should not be used for accurate work. Even with a 'standard' solvent, e.g. ethanol, extreme care should be taken to avoid changes in concentration due to evaporation. Solvents, e.g. benzene, ether, chloroform and carbon tetrachloride should also be avoided if possible as they are health hazards.

11.1.3 Interaction with the solute

The possibility of chemical reaction or complex formation with the solvent should be considered. Polar organic solvents nearly always cause shifts of spectral maxima, and in some cases these are very large. Similarly, changes in the pH of the solution can cause protonation or deprotonation of the solute giving rise to major changes in the absorption spectrum. When making comparisons with standard spectra, it is essential that the solvent composition, pH, ionic concentration, etc. are identical if valid conclusions are to be reached.

11.2 Making a solution

Having chosen a solvent, it is now necessary to decide upon a suitable concentration for the test solution. As a rule of thumb, for low-molecular weight organic molecules, 1 mg dissolved in 10 ml solvent should give a sufficiently concentrated solution for the measurement of the major maxima; e.g. if the molecular weight of a compound is 300 and the molar absorptivity of a typical peak is $2 \times 10^3 \, M^{-1} \, cm^{-1}$, then the solution in a 10 mm cell will give a peak absorbance of

$$A = \epsilon bc = 2 \times 10^3 \times 1 \times \frac{1 \times 10^{-3}}{300 \times 0.01} = 0.67$$

which is a suitable value for measurement.

The optimum absorbance range for a particular instrument is determined by a number of factors, which will be discussed in Section 11.7. The optimum range for most modern instruments is about 0.3–1.5 A. Older single-beam instruments using photocells rather than photomultipliers generally have a lower optimum range of 0.15–0.7 A. The critical parts of the measured spectrum should be kept within the appropriate limits.

The solute should be weighed out as accurately as possible. This may be considered unnecessary if the identification of the solute is the only purpose of running the spectrum; however, molar absorptivity values for the peaks may also be of help in identification or give some clues about the purity of the material. It is often necessary to repeat the determination at a later date and quantitative information then becomes valuable.

For the same reasons, the solute should be dissolved as completely as possible in an accurately defined volume of solution. These two aims cannot always be achieved in one operation! If you are confident that the compound will dissolve readily at room temperature, then the solution can be made up in a volumetric flask. The solid is put into the flask using a small glass funnel and the residue washed into the flask using about 80% of the final volume of solvent. The flask is stoppered, shaken until all the solute is dissolved and then made up to the mark. For unknown or less soluble material, the solution must be made before it is put into the volumetric flask. Use a beaker or conical flask, which can be stirred more conveniently and can be heated if necessary: never heat volumetric glassware for it may never return to its original volume. For granular material, a good technique is to put the solid in a small agate mortar, add a small amount of solvent and grind the solid to a paste. Further amounts of solvent are added and mixed in. The undissolved material is allowed to settle, the supernatant transferred to a volumetric flask using a Pasteur pipette and the procedure repeated until the solid is all dissolved.

Dissolution can be speeded by heating the solvent, but care should be taken that the solute is not affected by this. Often prolonged mechanical shaking or stirring at room temperature will succeed with apparently insoluble compounds. Immersion of the flask in an ultrasonic cleaning bath is an excellent means of speeding dissolution. If the weighed sample fails to dissolve completely, decant off the solution and add further solvent to the residue. The possibility that the original material was not homogeneous should be considered, particularly if the second batch of solvent fails to dissolve any more of it.

The success of the solution is usually judged by eye, but it is easy to be deceived by colourless compounds of refractive index near to that of the solvent. If there is any doubt, measure the spectrum, filter or centrifuge the solution, and measure again to see whether any change has occurred. If so, particles were present in the solution and its concentration is less than supposed.

11.3 The cell

The choice of cell for a particular task was dealt with in Chapter 2 and is, of course, related to the choice of solution concentration. The normal open-topped 10 mm cell is always first choice, for it is easy to handle, fill, empty and clean. They are, however, easily knocked over when standing on a bench and should always be stood in a small beaker or fitted into the cell-holder before being filled. The packaging for plastic cells can form a useful stand. All double-beam and some single-beam instruments need a second 'reference' cell containing 'solvent', i.e. all the components of the sample solution except the sample itself. The purpose of this cell is to compensate for the absorption, reflection and refraction of the sample cell and the solvent, and so the optical characteristics of the cells should match as closely as possible. Opinions vary about the necessary precision of this match. Traditionally, glass and natural quartz cells were sold in 'matched pairs', since the transmission of these materials varies considerably at short wavelengths. On the other hand, modern synthetic fused silica is effectively clear down to wavelengths below 200 nm, and such cells only need to be matched for the most precise work. The reference cell may seem redundant, particularly in modern instruments with built-in baseline correction, but the correct philosophy is to operate the instrument in the 'null' condition with the baseline as flat as possible. This becomes particularly important at short wavelengths, for in this region the transmission of the cell and solvent are changing rapidly, and the sensitivity of the system is falling. Conventional compensation by means of a reference cell is the best method, for it is effectively simultaneous with the sample measurement, for digital baseline correction is always subject to temporal effects, e.g. warming-up or drift of the instrument, or thermal effects on the cell or solvent. Some cells, e.g. flow cells, are too bulky or too expensive to allow an identical reference to be used. A compromise is to use an ordinary cell of the same window material and pathlength as the reference. If the sample cell has a reduced aperture, it is desirable to mask the reference beam in the same way.

Cells should never be touched with the fingers or handled by the optical faces. Dry cells can be handled with cotton gloves and wet ones with rubber gloves, but it is better to use plastic forceps, or metal tongs or forceps with their ends protected with short lengths of plastic tubing. The cell should be filled using a glass or plastic pipette. Pasteur pipettes can be used but their volume is too small for a normal 10 mm cell, and for the greatest speed and to avoid the risk of solution entering the bulb it is worth making special transfer pipettes, e.g. that shown in Fig. 11.1. Great care should be taken in filling the cell. The tip of a glass pipette should never be directed towards the windows for these are readily scratched, and for repetitive work it is advisable to protect the end of the pipette with a short length of plastic tube. Of course, when using organic solvents, the compatibility of plastic pipettes or tubing should be checked. It is now possible to obtain commercial supplies of polythene transfer pipettes which are free from UV-absorbing extratives.

Care must be taken when withdrawing the pipette to ensure that a drop of solution does not run down the outside of the cell. If this contaminates

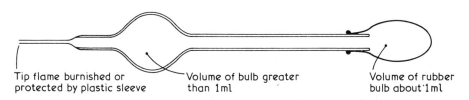

Fig. 11.1 A specially made transfer pipette for filling 10 mm normal rectangular cells [1].

the working area, the only cure when making measurements of the highest accuracy is to empty the cell, clean and dry it and fill it again. In routine circumstances, and if the spillage has not been allowed to evaporate on the cell window, it can be rinsed off with water or solvent and the cell allowed to dry. Again, good practice dictates that nothing should touch cell windows: this is imperative for cells used for high-accuracy work or when working in the far-UV. In less demanding situations, the cell can be mopped with lens tissue. Do not rub the cell windows and do not use ordinary laboratory tissues, for these contain abrasive substances and fluorescent brightening agents that will contaminate the windows. Be careful that grease from your fingers is not transferred on to the windows, particularly when using organic solvents, for such contamination can give appreciable absorption and can even cause the appearance of spurious peaks. A serious source of error in spectrometry is the filling and emptying of cells and their relocation in the cell-holder. For this reason, the recommended practice is to place the cell in the instrument at the beginning of a series of experiments and not to remove it until the last sample has been measured. If the cell is not removed from the instrument, extra care must be taken when filling and emptying it, and if a large number of samples is to be measured, a sipper or flow system will increase speed and accuracy while avoiding the risk of spillage inside the sample housing.

11.4 Making the measurement

Before measuring the sample, there are several routine operations that must be performed, these are neglected at your peril, for inquests on poor data can be far more time-consuming than these simple precaution.

11.4.1 Warm-up period

Modern instruments have solid-state electronics that warm up quickly, but tungsten lamps still need 10 min to come to thermal equilibrium and, particularly when old, can give rapid fluctuations during this period. Deuterium discharge lamps are arranged not to strike for 1–2 min after switching on and after this the arc will take some time to settle down. It is good practice to allow 20–30 minutes for stabilisation prior to making measurements.

11.4.2 Simple instrument checks

Ensure that the correct cell-holder is in the instrument, that it is properly located and all the cell positions are empty. Check that the correct lamp for the spectral range of interest is alight and that the lamp, filter and detector controls are correctly positioned. On recording instruments, check that there is sufficient paper, that the pens are working and that they are likely to continue to do so.

11.4.3 Instrumental baseline

This can be omitted if the instrument is in regular use over the spectral range to be used. However, the baseline is a useful check of the condition of the whole system and need not take long. Have the cell-holders that you intend to use in place, but ensure that there are no cells in them and that their apertures are free from dust and incrustations. Set the instrument operating conditions that will be used for the sample and scan fairly quickly over the range of interest. Even if measurements are to be made at a single wavelength, it is worth scanning on each side of that wavelength to ensure that the baseline is not changing rapidly. The baseline should be smooth, correspond to 1.0 T or 0 A and be free from excessive noise. If these conditions are not met, consult Section 11.5 or the more elaborate checks described in Chapter 9.

11.4.4 Cell baseline

This should be performed every time that clean cells are put into the instrument, and at least once per day. For the most accurate work, it is essential to run the cell baseline on each piece of chart paper that is used. The sample and reference cells are placed in position and filled with the appropriate solvent. When using cells with a restricted working area, e.g. micro-cells, check that the beam passes through the cell without touching the cell walls. This check can be done by turning the wavelength to 550 nm, opening the slit, and in subdued lighting or with a piece of black cloth over the open cell compartment, trace the path of the beam through the cell holder and cell by means of a small piece of white card. With micro-cells, some loss of light is inevitable, but it is essential that this falls on the mask and that there is no risk of it passing through the cell walls or over the sample. The best micro-cell-holders are provided with some means of adjusting the position of the cell laterally. Failing this, it may be possible to adjust the position of the cell-holder in the instrument so that the cell is centrally placed in the beam. If there is any risk of movement of the cell in the holder, it can be wedged against the reference edge or corner with pieces of card or thin metal.

The vertical alignment of the beam should also be checked with a known volume of water in the cell. The bottom of the beam should be above the cell floor and the top of the beam should fall at least 1 mm below the lowest part of the meniscus. If a limited amount of solution is available, the cell should be raised in the holder by packing pieces of card or plastic to minimize the clearance between the beam and the cell floor. If the sample volume is insufficient to fill the cell to the proper level, then the vertical height of the beam must be reduced with a mask.

Having positioned and filled the cell or cells with solvent, scan through the wavelength range. Since no peaks are expected in this trace, it can be scanned fairly rapidly. If the cells and the solvent they contain are identical, this trace should coincide with the instrumental baseline. In practice, this is rarely the case, particularly at short wavelengths. When approaching the short-wave solvent cut-off and the instrument is 'running out of energy' as a result, deviations from the expected baseline may result from the pen running off-scale or going 'dead'. Otherwise, if there are considerable differences from the instrumental baseline, check that the cells look clean and that there are no visible particles suspended in the solvent. Exchange the cells and see whether the deviation is reversed in sign. Empty and refill the cells with fresh solvent and re-measure. If the problem remains, clean the cells again and fill them with freshly distilled water. If they still show imbalance, this may be due to inherent differences in the transmission of the windows, contamination that your cleaning procedure has failed to remove, or fine scratches that are not visible when the cell is filled. If it is decided to proceed with the measurement using these cells, make regular checks of the solvent baseline to ensure that it is not changing during the day due, e.g. to the dissolution of contaminants from the windows.

11.4.5 Measuring the sample

The solvent is drawn out of the sample cell and, if sufficient sample solution is available, a few drops are used to rinse the cell. It is then filled to the correct level. It may not be possible to fill and empty short-pathlength cells while they are in the instrument. The tip of a standard Pasteur pipette is liable to jam inside a 1 mm normal cell and may scratch the windows or break off inside the cell. It is better to draw your own pipettes from tubing with a thicker wall so that the outside diameter is about 0.5 mm, and flame-burnish the end.

Before running the spectrum, a choice of operating conditions must be made. The number of operating parameters that can be chosen by the operator varies between instruments, but the aim should be to scan the selected wavelength range as fast as possible without any loss of accuracy of the record, both in wavelength and absorbance terms. If the scan is made faster than the data capture system can deal with, maxima will appear displaced to longer wavelengths, be reduced in height and distorted. Selection of the scanning speed is largely a matter of experience based on the performance of your instrument over a particular wavelength range and when measuring a spectrum of the same complexity. If the spectrum has sharp peaks with fine structure, a slow speed must be used. If you are uncertain about the accuracy of the resulting spectrum, scan it again at a slower speed and compare the results. If the peaks are unchanged in position and shape, the original speed was satisfactory. Figure 11.2 shows the result of scanning the benzene vapour spectrum, which has very sharp peaks and more fine structure than is seen in most solution spectra, at different speeds with a conventional recording spectrometer. An ideal solution to the scanning speed problem is presented in some modern microprocessor-controlled spectrometers where the scanning speed is automatically regulated by the rate of change of the absorbance, the instrument scanning fast over flat regions of the spectrum and slowly over sharp peaks.

While most instruments allow a choice of recording speed, far fewer permit a choice of slitwidth (ESW). Here the compromise is to select the smallest ESW and hence greatest spectral resolution consistent with reasonable sensitivity. In prism instruments, the ESW for a given slit setting varies with wavelength due to the changing dispersion of quartz. This is illustrated in Fig. 11.3. When using manual instruments with quartz prisms, the operator must select a slitwidth for each wavelength, the choice being based on the ESW, the lamp output and the detector response at that wavelength. The manufacturer's manual generally gives the slit settings to maintain constant sensitivity. In most recording prism instruments, the slit servo operates to maintain a constant response from the detector and hence a roughly constant sensitivity at all wavelengths. This means that there are considerable fluctuations in the ESW as the instrument scans, as shown by the broken line in Fig. 11.4. In critical applications, in order to obtain a specific ESW at a given wavelength, it may be necessary to override the slit mechanism and set the slit by hand.

The problem is simpler in grating instruments, for in these the dispersion is constant and so the instrument has constant ESW with a fixed slitwidth. It is still necessary for the instrument to compensate for changes in the lamp output and detector sensitivity with wavelength, but this is usually done with an optical wedge or comb that adjusts the amount of radiation without changing the ESW.

As a rule-of-thumb, the ESW should not be more than the width at half-height of the sharpest band to be measured or, alternatively, the ESW should be less than one-eighth of the bandwidth: thus for a peak of bandwidth 40 nm, the ESW should not exceed 5 nm. Figure 11.5 demonstrates the effect of bandwidth on the apparent height of a sharp peak in the benzene spectrum, but significant deviations can occur with relatively broad peaks if excessive ESW settings are used. If you are uncertain about your choice of ESW, scan the critical peak again with a smaller slit setting.

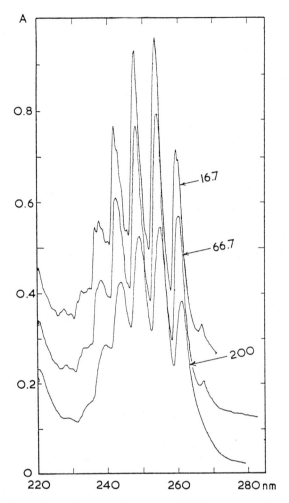

Fig. 11.2 Effect of scanning speed on a recorded spectrum. A cell containing benzene vapour was scanned from short to long wavelength at three different rates given as nm min^{-1} on a Shimadzu MPS-5OL double-beam spectrometer. Excessive speed has reduced the fine structure, lowered the peaks and raised the minima, and has shifted both maxima and minima to longer wavelengths. These effects are due primarily to the slowness of the pen response.

Even if the slitwidth of your instrument cannot be changed, it is essential to know the ESW at different wavelengths and appreciate that a potential problem exists.

The instrument is then set to the starting wavelength by driving the mechanism in the direction of the intended scan so that the backlash in the driving mechanism is taken up. If possible, the pen should be allowed to equilibrate before starting the scan to avoid generating spurious peaks. If

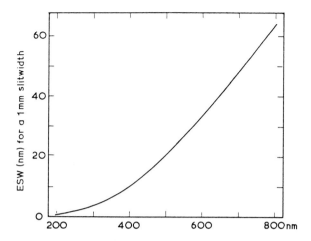

Fig. 11.3 A typical dispersion curve for a quartz prism monochromator. This shows the effective spectral slitwidth at different wavelengths for a slit opening of 1 nm.

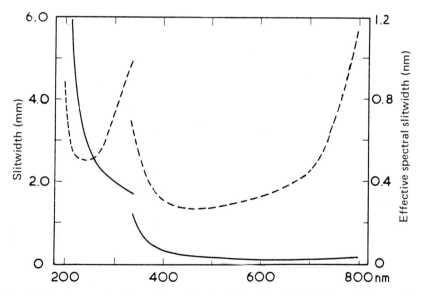

Fig. 11.4 Automatic slit settings for the Shimadzu MPS-5OL under normal operating conditions (continuous line) and the resulting ESW values calculated from Fig. 11.3 (broken line).

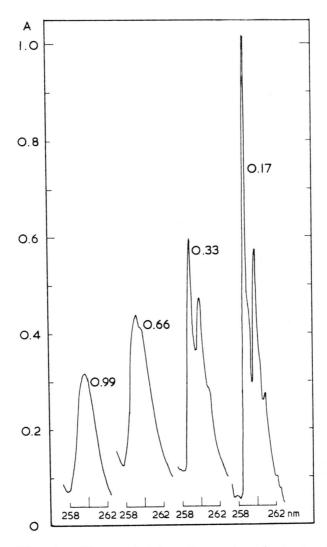

Fig. 11.5 Distortion of the recorded absorption spectrum due to excessive ESW. A cell containing benzene vapour was scanned at the same rate with four different ESW values, which are given in nm. The spectrum at the smallest slit setting approaches the true shape of the band, the peaks of which are extremely narrow.

the peak absorbance of the sample lies outside the chosen absorbance range, it is generally better to complete the scan and then go back and fill in the missing parts of the spectrum on another scale. If the critical part of the spectrum is found to lie above the recommended range, i.e. 1.5 A, consider diluting the sample or changing to a shorter pathlength cell.

At the end of a series of measurements, the cells should be removed from the instrument, emptied and submerged in solvent or one of the cleaning solutions suggested in Section 11.6. Cells should never be allowed to dry out, as this makes cleaning far more difficult.

11.5 Problems and pitfalls

Experience will teach when a spectrum does not 'look right' even though the above procedures have been carried out. Some commonplace problems will be described.

11.5.1 Poor instrumental baseline

Check that there is nothing in any cell-holder, that they are correctly aligned and, if the instrument has manually operated filters, shutters or attenuators, check that these are correctly positioned. Check that the appropriate lamp is alight, and if the change over is manual, that the lever is properly set. Check that the beam is passing correctly though the cell-holders by turning to 550 nm, opening the slits as wide as possible and tracing the beam with a piece of white card. If the instrument has the means of displaying the energy level in single-beam mode, check this: otherwise, excessive noise on the trace suggests that lamp output is low or that there is failure of an electronic component. If the instrument has some means of compensating the baseline at different wavelengths, this may need resetting, particularly if there has been a change in ambient temperature or the instrument has not been used for some time. If the baseline has sharp spikes or steps that always occur at the same wavelength, suspect some fault in the automatic filter-changing system. Further checks are given in Chapter 9.

11.5.2 Poor cell baseline

If the instrumental baseline is satisfactory, then the fault must lie with the cells or their contents. Empty and refill the cells to ensure that the solvent is homogeneous. Visually check the cells for dirt inside and for streaks or scratches on the windows. If using solvents other than pure water, empty the cells and refill with water, for high-solvent absorption at short wavelengths may show up a latent imbalance in the instrument or differing stray-light levels in the two beams. If the problem is still not resolved, try another pair of cells or clean the cells again using a more drastic method.

11.5.3 Low-absorbance values

When measuring the sample, the absorption maxima may appear lower than expected for a number of reasons:

(a) Incomplete solution – check that there is no residual material in the flask and no suspended particles in the solution. Check that the

solute has not come out of solution in the cell as a precipitate or as a coating on the walls.
(b) Sample impurity – it could be contaminated with a non-absorbing impurity.
(c) Changes in the sample – check that the pH of the solution is correct. Could the sample have suffered photochemical or oxidative damage? Check the rest of the spectrum to see whether there has been a corresponding absorbance increase due to product formation, and run the spectrum again to see whether the change is a progressive one.
(d) Bubbles in sample or reference cells – if there have been temperature changes or the cells have been filled for some time, bubbles may have formed on the windows.
(e) Light by-passing the sample – check that the beam cannot pass above, below or either side of the sample, and that there are no bubbles in the cell.
(f) Peak distortion due to excessive scanning speed or too great an ESW – this is only likely to occur with very sharp peaks.
(g) When making measurements at short wavelengths always be alert to the contribution of stray-light. This is equivalent to light by-passing the sample and so will reduce the apparent absorbance (see Fig. 11.6). The problems can be improved by diluting the sample or using a shorter pathlength cell so that the light transmitted by the sample is increased and the stray-light is a smaller proportion of the light reaching the detector.
(h) Beer's law failure – the linear relationship of absorbance with concentration may appear to fail as a result of a change in the molecular species in solution due to dimerization, complex formation, etc. If such a problem is suspected, measure a series of dilutions and see whether the change is progressive. While Beer's law failure is often invoked to explain non-linearity, in the majority of cases the error has some other cause.
(i) Fluorescence – there is a possibility that the measuring beam may cause the sample or solvent to fluoresce: if this emission enters the detector it will reduce the apparent absorbance. This is unlikely to cause serious error.

If none of the above suggestions seems to explain your problem, it will be necessary to make a proper check of the absorbance accuracy of the instrument using the methods detailed in Chapter 13.

11.5.4 High absorbances

This is a far less common problem, and so in double-beam instruments it is worth first considering whether the transmission of the reference beam

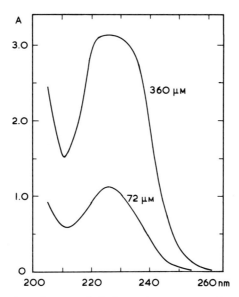

Fig. 11.6 Distortion of a recorded absorption spectrum due to stray-light. Two aqueous solutions of potassium iodide differing in concentration by a factor of 5 were measured in a 10 mm cell on a Pye Unicam SP8–100. The curve at low concentration represents the true spectrum, while that for the high concentration is distorted below 240 nm. The apparent λ_{max} is shifted and is only about 60% of its true height.

has increased for any reason. Otherwise, the fault is probably optical obstruction of the sample beam by bubbles, evaporation of the solution increasing its absorbance or causing the meniscus to descend into the beam, or the formation of a precipitate in the solution. When making measurements at temperatures below ambient, be alert to the possibility of condensation forming on the cell windows, the solute coming out of solution, or the solution freezing, all of which will increase the apparent absorbance.

11.5.5 Incorrect λ_{max} values

Check that the wavelength scale on the chart corresponds to that on the monochromator. In critical applications superimpose the transmission profile of a calibrating filter or solution on the same chart: remember that chart paper will expand with increases in humidity as the day goes on, and the wavelength calibration of the instrument can drift as it warms up.

When measuring mixtures, the apparent λ_{max} of minor components appearing on the side of major peaks will be displaced from the true value. The effect of the superposition of a minor peak on the tail of a major one is illustrated in Fig. 11.7(a). If there is overlap of two peaks of similar size,

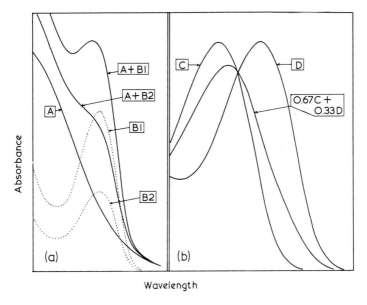

Fig. 11.7 Diagram illustrating the distortion of spectra by the overlap of bands: (a) the apparent λ_{max} of the minor band is displaced towards the maximum of the major peak; (b) a 2:1 mixture of compounds with overlapping spectra results in a single peak of intermediate λ_{max}.

the resulting single peak gives no clue as to the presence of two components [Fig. 11.7(b)]. Changes of pH or solvent can give major differences in λ_{max} from published values and, when this information is available, it is worth checking that the concentration and temperature of the solution are similar to those of the reference. As we have seen, stray-light can cause spurious peaks to appear at short wavelengths and can also displace the position of true peaks. If the peaks are flattened and have an asymmetrical appearance, stray-light is probably present.

11.5.6 Spurious peaks

Unexpected peaks that are not present in the solvent baseline can be the result of impurities or degradation products in the sample or due to contamination introduced during the handling of the solution.

11.6 Cell cleaning

Cell cleaning procedures were dealt with in Volume 1 of this series [1] and will only be briefly dealt with here. The first principle is to avoid drastic cleaning methods if possible, for they are time consuming and potentially dangerous to both cell and user. Many cleaning problems can be avoided by not leaving solutions in cells for longer than necessary and never allowing them to evaporate.

When cleaning a cell, first assess the problem. Is there a suitable solvent for the contaminant that will not damage the cell? Plastic cells can only be cleaned with dilute solutions of detergents. Glass cells must not be treated with alkaline solutions, and there is a possibility that cheaper cells may be of cemented construction which will exclude the use of strong acids and alkalis and some organic solvents. Silica cells of fused construction are effectively inert, and can if necessary be heated in the cleaning solution.

Organic contaminants may well be dissolved by a suitable organic solvent. Failing this, a cold or warm solution of a detergent may be successful. The next approach is to oxidize the offending substance with chromic acid mixture or hot nitric acid. Extreme care must be taken when handling these reagents: this should be done in a fume hood wearing gloves, goggles and protective clothing. If oxidizing agents are used, it is essential to make sure that the cell is as free of organic matter as possible; first rinse it with acetone or ethanol and then several times with water.

Finally, the cell should be thoroughly rinsed with distilled water. If it is in regular use with aqueous solutions, it is best to store it submerged in water, upright in a narrow container so that it cannot fall over. If it is necessary to dry the cell, empty it and stand it inverted on an absorbent surface. A small plastic-coated test-tube rack can be used to prevent the cell falling over, but this must not touch the windows. To dry the cell more quickly, it may be rinsed with spectroscopic grade alcohol or acetone; this should be drained from the cell as completely as possible before allowing it to evaporate.

Cells should never be touched with the fingers, especially when wet. When a routine of good cell handling has been established, it takes no longer than more casual methods, guarantees the best possible results and avoids time-wasting inquests on poor results.

Dry cells should be stored in small sealed boxes free of dust and lint. Plastic sandwich boxes are ideal, but the plush-lined boxes supplied by some manufacturers should be discarded. For transportation or long-term storage the cells may be wrapped in lens tissue.

11.7 Accuracy and precision in absorbance measurement

Your sample has been measured with as much care as possible and your instrument has produced an absorbance value at a particular wavelength. How valid is this number? Any measuring system has two criteria of performance, precision and accuracy. These are illustrated in Fig. 11.8 by the performance of a marksman in a rifle range. In (a) the shots are widely distributed all over the target, this shows neither precision nor accuracy. In

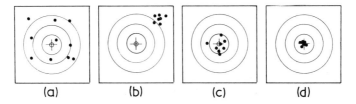

Fig. 11.8 The concepts of precision and accuracy are illustrated by the performance of a marksman firing at a target: (a) lacks both precision and accuracy; (b) the hits are precisely located but are inaccurate; (c) here the accuracy is good but the precision poor; (d) shows both precision and accuracy. From Ref. [2].

(b) they are grouped together away from the bull's-eye: the marksman and rifle are performing well and giving a precise distribution of shots, but there is a bias, perhaps the rifle sight is bent. Target (c) shows a grouping with satisfactory accuracy but poor precision: the average of a large number of hits would lie on the bull's eye. In (d) the hits are both precise and accurate. The analogy with absorbance measurement is good: both instrument and operator may be working reliably to give precise data and repeated measurements on the same sample show small deviation, but some error in sample preparation or instrumental bias means that the mean of the measurements does not represent the true absorbance. To obtain results that are both precise and accurate demands precision and accuracy from both operator and instrument. Precision is checked quite easily by repeated measurements. Modern instruments are capable of extremely high precision and in 99% of cases it is operator error that leads to poor data. On the other hand, assessment of accuracy is far more difficult, and in laboratories where quantitative results are vital, a great deal of effort must be put into checking and improving the accuracy of absorbance measurements.

11.7.1 Tests for precision

Suppose that you are involved in the routine measurement of a number of solutions of the same solute. The first test would be to take a typical solution and perform your normal routine of filling, measuring and emptying the cell, e.g. 20 times. Calculate the standard error of these results:

$$s = \left(\frac{\Sigma x^2 - (\Sigma x)^2 / n}{n - 1} \right)^{1/2} \tag{11.1}$$

The standard error will depend upon the instrument and the nature of the sample, but should be of the order of 0.005 and not exceed 0.02. If it does, consider the following potential causes of deviation: (a) inhomogeneity of sample; (b) contamination of the external faces of the cell by spillage or fingering; (c) contamination internally by the solute; (d) movement of the cell; (e) instrumental instability; (f) solvent carry-over, if the cell is rinsed between measurements, the sample may be diluted by residual solvent. If the test is satisfactory, next assess the performance of the system over the anticipated range of absorbance readings.

11.7.2 Linearity checks

Prepare a series of dilutions of the most concentrated sample and measure their absorbance using your normal routine. Plot these values against concentration. Ideally it should be linear over the whole range with the points all falling on the line. The line may deviate from linearity due to stray-light, instrumental non-linearity or Beer's law failure, but most probably due to poor volumetric technique. Sample handling errors can probably be diagnosed by repetitions of the measuring procedure, but instrumental and Beer's law problems are difficult to distinguish. If a solution is measured in cells of different pathlength, then any non-linearity in the absorbance–pathlength relationship must arise in the instrument. On the other hand, if solutions of differing absorbance are measured in the same cell, any non-linearity may be due to Beer's law failure, instrumental non-linearity, stray-light, or some combination of these effects. As Clarke [3] points out, a test of this sort may underestimate the Beer's law deviation.

Scatter of the points may be due to any of the causes outlined in the previous section. If the scatter becomes worse at high or low absorbances, then it may be instrumental fluctuations that are limiting the precision of the results. In any case, it is prudent to consider in what ways the precision of the measurements can be improved.

11.7.3 Optimization procedures

Having established that the sample handling is being performed as carefully as possible and that the instrument is working well, precision can be further improved in two ways: first, by increasing the number of measurements and thus reducing the standard error, and second, by ensuring that measurements are made within the optimum range of the instrument. Repeated measurements will only increase precision if all the critical steps in the measurement are repeated: if cell filling is the prime cause of deviation, then the cell must be filled and emptied for each of the replicates. Selection of an optimum range is more concerned with the performance of the instrument. Figure 11.9 shows the instrumental error in the measurement of a series of solutions of a given solute at a specific wavelength on a particular instrument. The curves represent the

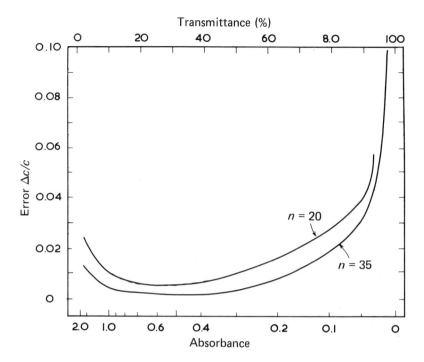

Fig. 11.9 Measurement of instrumental error in a spectrometer. The transmission of a series of 12 solutions of potassium chromate measured at 370 nm with a Bausch and Lomb Spectronic 20. Each solution was measured 20 and 35 times, and error values based on the deviation of these points were calculated by Equation (11.2). From Ref. [7].

deviations in the absorbance values when each of the solutions was measured a number of times. The 0 and 1.0 T of the instrument was reset before each measurement, but the sample cell was left in the cell holder. As expected, the precision falls at high and low absorbancies, while the precision is increased for the number of repeated measurements. The optimum absorbance range for a given level of precision can thus be increased by increasing the number of repetitions.

There are a number of intrinsic causes of instrumental deviation, the most important being noise generated by the detectors ('shot' noise), noise generated by the electronic components of the amplifiers, fluctuations of the source and 'noise' generated by the uncertainty of the read-out. The contribution of these various kinds of noise to the signal-to-noise ratio is discussed in detail by Rothman *et al.* [4], who make recommendations on the selection of operating parameters to maximize the signal-to-noise ratio.

An early attempt to establish the optimum absorbance range of the Beckman DU spectrometer was made by Vandenbelt *et al.* [5] who measured the molar absorptivities of various compounds at different concentrations. Three typical curves are given in Fig. 11.10, which shows the deviations of the apparent molar absorptivity when measured with solutions of high and slow absorbance. The linear part of the plot indicates the absorbance range of greatest accuracy. Interestingly, all three curves are of the same shape, although they were measured at different wavelengths and correspond to widely differing solute concentrations. The downward trend in molar absorptivity values above 2 A is probably due to stray-light, but the sharp rise in the values at low absorbance is less readily explained. The authors suggest that it may have been due to photometric non-linearity in the system.

An alternative approach to the determination of the optimum concentration range for a spectrometric assay was proposed by Ayres [6] based on an earlier proposal by Ringbom. Rather than plotting absorbance against concentration for a series of test samples, transmission

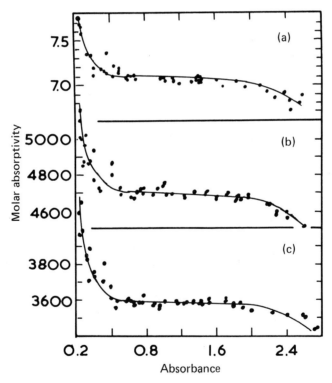

Fig. 11.10 Non-linearity in a manual spectrophotometer. A plot of the variation in the apparent molar absorptivity of two compounds at three wavelengths with absorbance of the test solution. (a) Potassium nitrate in water at 301 nm; (b) potassium chromate in 50 mM potassium hydroxide at 373 nm; and (c) potassium chromate in 50 mM potassium hydroxide at 273 nm. Redrawn from Vandenbelt *et al.* [5].

is plotted against the logarithm of their concentration (Fig. 11.11). The linear region of maximum slope indicates the range of concentration that will give maximum sensitivity and linearity for the determination. The method is based on the assumption that instrumental error in making transmission measurements is constant throughout the transmittance range, which is not true. A more rigorous method for estimating the region of maximum precision of an instrument is given by Youmans and Brown [7]. The error in the measurement of the concentration of a solution of transmission T can be expressed as:

$$\Delta c = \frac{1}{\epsilon b} \log \frac{T_{av}}{T_{min}} \tag{11.2}$$

where ϵ is the molar absorptivity of the compound, b the pathlength, and T_{av} and T_{min} are the mean and minimum values from a series of n measurements on the same sample.

The statistical error is then:

$$\frac{\Delta c}{c} = \frac{1}{-\log T_{av}} \log \frac{T_{av}}{T_{min}}$$

The standard error s of T_{av} for the population of n measurements can be calculated from Equation (11.1) and corrected for bias with Student's t-value. Then

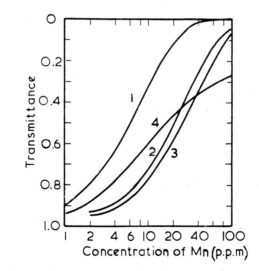

Fig. 11.11 An Ayres plot for the determination of manganese as permanganate. Curves 1–3 were obtained with a series of solutions measured at 526, 480 and 580 nm, respectively, using a Beckman DU manual spectrometer. Curve 4 was obtained using a filter colorimeter. Instrumental accuracy is greatest where the slope of the curve is greatest, deviations occurring at high and low transmittances. Redrawn from Ayres [6].

$$T_{min} = T_{av} - \frac{ts}{n}$$

In this way, the curves shown in Fig. 11.9 were calculated from groups of 20 and 35 measurements on each of a series of 12 solutions covering the transmission range. Values of T_{av} and s were calculated, and then T_{min} derived using Student's t-values taken from tables taking the degree of freedom as $(n-1)$ and $\alpha = 0.05$.

These curves, of course, refer to one particular type of estimation on a specific instrument, but are fairly typical. They show the relative intrinsic error in the system at different absorbance values. They do not form any basis for the correction of such errors but are of great help in the optimization of a particular assay.

11.8 Difficult samples

11.8.1 Solid samples

Satisfactory absorption measurements can be made on samples that are insoluble or must not be dissolved, though the techniques are more difficult and the results more liable to mis-interpretation. For large optically homogeneous specimens, e.g. single crystals, glasses or polymer sheets and blocks, satisfactory measurements can be made simply by fixing the sample in the measuring beam, with suitable masks if necessary. For the best results, the sample should have parallel plane faces and be arranged so that these are perpendicular to the beam. For irregularly shaped specimens, immersion in a liquid of matching refractive index can reduce the effects of surface reflection and refraction. Of course, the liquid must not absorb appreciably in the region of interest.

Smaller crystals can be measured with a micro-beam set-up (see Section 11.8.3) or by packing a quantity of them into a cell. Light scattering from the surfaces of the crystals will be a major problem, though the effects of this can be reduced (see Section 11.8.2). It can also be reduced by suspending the sample in a high-refractive index liquid to form a 'mull'. Everett [8] has measured insoluble compounds in the UV by mixing them with a non-absorbing salt and compressing the mixture into discs, just as in the IR 'pressed disc' technique. Light scatter is a major problem but can be minimized by optimizing the size of the salt granules.

For low-melting point solids, it is worth considering measuring them in the molten state: temperatures up to 100°C can be maintained in a suitably insulated cell-holder in a conventional spectrometer. The results should be interpreted with caution, for the liquid state spectrum may not be identical with that in the solid state or in solution. An alternative technique, which is the only one that can be used for highly absorbing solids, is reflectance spectrometry.

11.8.2 Scattering samples

A common problem, particularly with biological samples, is that the substance of interest is associated with suspended particles of other material. Although the latter may not absorb in the spectral region of interest, scattering of the measuring beam will distort the measured spectrum and may raise the apparent absorbance of the sample above the optimum range of the instrument. The scattering of radiation depends upon the size of the particles involved and is also a power function of the frequency of the radiation, i.e. its effect increases rapidly when scanning down into the UV region. This is illustrated by Fig. 11.12 which shows the effect of increased turbidity upon an absorption spectrum. Scatter increases the apparent absorbance, particularly at shorter wavelengths, and so the spectrum is distorted and its peaks are shifted to shorter wavelengths. Keilin and Hartree [9] presented a mathematical treatment of this effect.

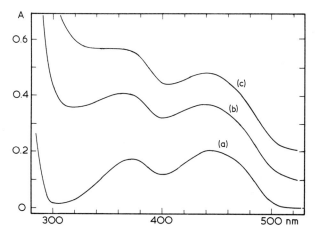

Fig. 11.12 The effect of light scattering by the sample, and instrumental compensation: (a) absorption spectrum of 16.9 µM riboflavin in water at pH 5.6 measured in a 10 mm cell with a Pye Unicam SP8-100; (b) the same solution with 1% milk added measured in the 'turbid sample' position, close to the detector; (c) the solution containing milk measured in the normal cell position.

Some degree of compensation can be made by using a scattering suspension in the reference beam, diluted milk or 'Ludox' colloidal silica may be helpful. Goldbloom and Brown [10] give an example of this. A better solution is to place the sample cell nearer to the detector, the latter then receives a greater proportion of the scattered radiation (Fig. 11.13) and the measured spectrum will be nearer the true shape. Some modern instruments have an alternative cell-holder placed in front of the detector to facilitate this. An alternative method for use with conventional

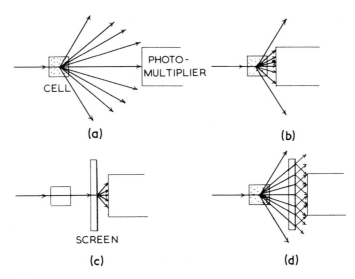

Fig. 11.13 Instrumental methods for correcting for light scattering. (a) Conventional arrangement of cell and detector. A scattering sample in the cell causes much of the radiation leaving the cell to miss the detector; (b) if the cell is placed close to the detector, a larger proportion of the radiation will enter it; (c) if a diffusing screen is placed between sample and detector, light passing straight through the sample will be scattered before entering the detector; (d) a scattering sample will show the same transmittance as a clear one since the same proportion of light will enter the detector irrespective of how it strikes the screen.

spectrometers was proposed by Shibata *et al.* [11] who showed that if a scattering 'screen' is placed between the cell and the detector, the absorption spectrum will be nearer its true shape. All light rays emerging from the cell will be scattered equally by the screen irrespective of their direction, and so a constant fraction of the light striking the screen will enter the detector. Shibata *et al.* used oiled paper as a screen, but 'flashed' opal glass is more satisfactory.

11.8.3 Microscopic samples

The measurement of microscopic samples is becoming commonplace, and many special instruments have been built for the measurement of forensic samples, chromosomes in cells, etc. Accessories are available for conventional spectrometers that will reduce the beam to microscopic size, but probably custom-built instruments based on a microscope are more satisfactory. Commercial instruments of this type are reviewed by Altman [2] and can be very expensive. Simpler home-built instruments designed for a particular task can give equally good results at far less cost. The instrument described by Liebman and Entine [12] is still in use in an updated form and produces excellent absorption spectra in the range 320–700 nm from specimens 0.5 × 2 μm in size. The major problem in making quantitative measurements with this type of instrument is to provide a satisfactory reference. In Liebman's instrument, a reference beam is taken through the microscope and focused on the specimen slide beside the sample beam. The operator manoeuvres the slide so that the sample beam passes through the sample while the reference passes through the mounting medium.

References

1. Burgess, C. and Knowles, A. (eds) (1981) *Techniques in Visible and Ultraviolent Spectrometry, Vol. 1, Standards in Absorption Spectrometry*. Chapman and Hall, London.
2. Altman, F.P. (1981) *UV Spectrom. Grp Bull.*, **9**, 28.
3. Clarke, F.J.J. (1981) *UV Spectrom. Grp Bull.*, **9**, 81.
4. Rothman, L.D., Crouch, S.R. and Ingle, J.D. (1975) *Anal. Chem.*, **47**, 1226.
5. Vandenbelt, J.M., Forsyth, J. and Garrett, A. (1945) *Anal. Chem.*, **17**, 235.
6. Ayres, G. (1949) *Anal. Chem.*, **21**, 652.
7. Youmans, H.L. and Brown, V.H. (1976) *Anal. Chem.*, **48**, 1152.
8. Everett, A.J. (1968) Personal communication.
9. Keirin, D. and Hartree, E.F. (1958) *Biochim. Biophys. Acta*, **27**, 173.
10. Goldbloom, D.E. and Brown, W.D. (1966) *Biochim. Biophys. Acta*, **112**, 584.
11. Shibata, K., Benson, A.A. and Calvin, M. (1954) *Biochim. Biophys. Acta*, **15**, 461.
12. Liebman, P.A. and Entine, G. (1964), *J. Opt. Soc. Amer.*, **54**, 1451.

12 Numerical Methods of Data Analysis

12.1 Baseline corrections

12.1.1 Irrelevant absorption

The measured absorbance at any wavelength is the sum of the absorbances of the various components present and this is the basis of the quantitative analytical methods, discussed in Section 12.2. This approach presupposes that all of the absorbing components present are known, or can be identified. However, this is frequently not the case, particularly with measurements on biological materials. There is background absorption, commonly termed irrelevant absorption, which is usually rather non-specific in that it lacks discrete maxima. If this absorption can be quantified and subtracted, the spectra may be used with confidence for assay work. Various methods have been described for the elimination of this irrelevant absorption, some very simple and others more mathematically demanding. The selection of a particular method is very dependent on the nature of the irrelevant absorption; e.g. if it is linear, two points, in principle, serve to define it. The earlier work on baseline correction procedures has been well reviewed by Mulder *et al.* [1], whose paper may be consulted to supplement the basic principles given below.

12.1.2 The Morton and Stubbs correction

The simplest approach to background absorption is to assume that it increases linearly below the peak of the absorption band being studied at the same rate as on the long wavelength side of the band. The weakness of this approach is that it usually involves a long extrapolation of the baseline and the assumption of linearity is then often not valid. It is therefore better to use a procedure which fixes the baseline over a short distance only on either side of the absorption maximum; this is the basis of the Morton and Stubbs method [2], which is demonstrated by Fig. 12.1. The upper curve is the measured absorption band, the lower broken curve is the true contour of the compound to be estimated and the background is assumed to be

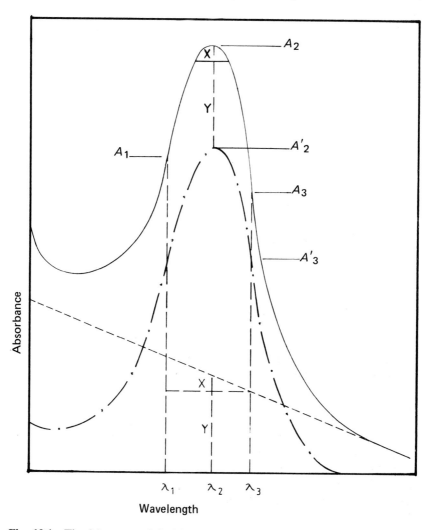

Fig. 12.1 The Morton and Stubbs baseline correction. A_1 A_2 and A_3 are the absorbances at three points on the observed absorption spectrum (——). A'_2 is the absorbance at λ_2 on the true absorption spectrum (–•–•–) after correcting for the background absorption (– – –) which is assumed to be linear between λ_1 and λ_3. Redrawn from Morton and Stubbs [2].

linear over the small wavelength range under consideration. Three wavelengths, λ_1, λ_2 and λ_3 are chosen, such that $\lambda_2 = \lambda_{max}$ and λ_1 and λ_3 have equal molar absorptivities, i.e. $\epsilon_1 = \epsilon_3$, which are a convenient, known fraction, B, of the value $\epsilon_2 = \epsilon_{max}$ at the band maximum. The measured absorbances at λ_1, λ_2 and λ_3 are designated A_1, A_2 and A_3, respectively, and the difference between A'_2 (the true absorbance at λ_2) and A_2 is the background absorbance $(x + y)$. Thus

$$A'_2 = A_2 - (x+y)$$

where

$$x = (A_1 - A_3)(\lambda_3 - \lambda_2)/(\lambda_3 - \lambda_1)$$

and

$$y = (BA_3 - (A_2 - x))/(B - 1)$$

The wavelengths λ_1 and λ_3 must be chosen with some care. If $\lambda_3 - \lambda_1$ is large, the assumption of a linear background over the interval is likely to be invalid. However, if λ_1 and λ_3 are very close, the absorbance values at these two wavelengths and at λ_2 will be rather similar, so that the differences between them may be comparable with the instrumental error. Morton and Stubbs developed their procedure specifically for the estimation of vitamin A in fish liver oils and, in these particular circumstances, chose λ_1 and λ_3 such that ϵ_1 and ϵ_3 were both 6/7 of the value of ϵ_2, giving a wavelength interval $\lambda_3 - \lambda_1$ of about 25 nm. The precision of the method, in the context of the estimation of vitamin A, has been examined in detail by Adamson et al. [3], who give estimates of the standard deviations likely to be encountered in careful spectrometric work.

The estimation of ergosterol, using the band at 282 nm, proves more difficult because the band is very sharp; a wavelength setting error of only 0.5 nm with λ_1 and λ_3 gives absorbance, and hence concentration, errors of about 20%. This led Shaw and Jefferies [4] to modify the Morton and Stubbs method, in that they selected λ_1 and λ_3 to coincide with adjacent maxima in the ergosterol spectrum. Although Morton and Stubbs [5] used the particular condition $\epsilon_1 = \epsilon_3$, it is possible to apply the method in the more general case $\epsilon_1 \neq \epsilon_3$, as they and Shaw and Jefferies have pointed out. Although the relevant equations are more complex they are not difficult to use.

12.1.3 Non-linear irrelevant absorption

Not surprisingly, it is necessary to have methods available for dealing with non-linear irrelevant absorption, for those situations where there is non-linearity over a rather narrow wavelength range. Such non-linear backgrounds may be fitted to equations of the type

$$A = a + b\lambda + c\lambda^2 + d\lambda^3 + \ldots$$

The characterization of such backgrounds requires absorbance measurement at not less than four wavelengths. However, the problem with these non-linear backgrounds is that they are often not of constant shape among a series of samples, as Ashton and Tootill [6] observed in the assay of griseofulvin. There is, therefore, the need for a different approach and this is provided by the use of harmonic analysis.

The basis of the method is that a given function, in this instance the absorption peak of the compound being estimated, can be expressed as the sum of a set of fundamental shapes. This is expressed by the equation:

$$f(\lambda) = ag_0 + bg_1 + cg_2 + dg_3 + \ldots$$

where g_0 is a constant, but g_1, g_2, g_3, etc. are functions of λ, and a, b, c, d, etc. are coefficients. In harmonic analysis, g_1, g_2, g_3, etc. are trigonometric functions, but Ashton and Tootill [7], and Glenn [8] have shown that Legendre polynomials are more convenient for computational purposes. These are also known as orthogonal polynomials. An absorption band is then expressed in the form

$$f(\lambda) = aP_0 + bP_1 + cP_2 + dP_3 + \ldots$$

The first six Legendre polynomials are shown in Fig. 12.2, and it is evident that any peak may be represented as the sum of a constant background, a linear slope factor and a series of non-linear terms. If the irrelevant background absorption is expressed in a similar way, P_0 and P_1 terms will be present, together with some P_2 if the background is of quadratic form, but it is unlikely that the coefficient of P_3 will be appreciable. Hence, the coefficient in P_3 of the measured absorption curve is due solely to the component to be estimated and contains no contribution from the baseline. Furthermore, it is directly proportional to its concentration and is used in place of the more usual ϵ. The procedure, therefore, is to express the measured absorption curve in terms of Legendre polynomials and to use one of these, probably P_3, for the determination. The simple computational details have been set out lucidly by Glenn [8]. The role of Legendre polynomials has been discussed in relation to wavelength calibration by Clewes [9].

12.2 Data smoothing

12.2.1 Noise and smoothing

Random fluctuations, usually referred to as noise, occur in all spectrometric measurements. They arise largely, but not exclusively, in the detector system and associated electronics. This noise limits the precision that may be achieved in quantitative work and also sets the limit of detection of minor components. Electrical smoothing circuits are included in spectrometer detector and output systems, but it is also advantageous to have the facility for mathematical smoothing. Not only does it provide

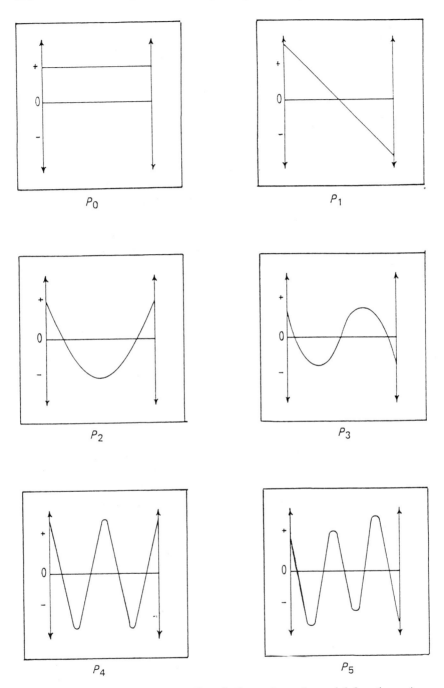

Fig. 12.2 Representation of the first six Legendre polynomial functions. Any spectrum can be resolved into a sum of such functions. Redrawn from Clewes [9].

greater flexibility, it operates on a fundamentally different basis to instrumental smoothing. The latter can only process the information that has already passed through the system, whereas mathematical smoothing on data collected in digital form provides the means for involving data points both before and after the one under consideration. Two methods are in general use for digital smoothing, the moving average and least squares methods.

12.2.2 The moving average method

Because noise is usually random in character, the averaging of a series of data points symmetrically placed about the one under consideration will lead to a reduction in the noise level. If the spectrometer output, e.g. absorbance, at three consecutive and equally separated wavelengths has the values A_1, A_2 and A_3, the simplest average is $(A_1+A_2+A_3)/3$. Alternatively, it may be useful to give more weight to the value of interest, A_2, by using an expression $(A_1+2A_2+A_3)/4$. The smoothing operation consists of calculating such a three-point average for λ_1, λ_2 and λ_3, then moving on to λ_2, λ_3, λ_4 until the last three points λ_{n-2}, λ_{n-1} and λ_n of the data set are reached. This is a very simple computational procedure and it is possible to use more than three points in the data set. However, this method assumes that there is linearity between A and λ over the interval used, and if the wavelength interval is too large there will be an appreciable deviation and distortion.

Trott and Beynon [10] have recently extended the simple three-point moving average method, in two respects. They showed that if the process is repeated on the same data set there is a further improvement and, in principle, this should continue up to N passes, where $N=0.5n-1$ for n even and $0.5(n+1)-1$ for n odd, and n is the number of points in the data set. However, they showed experimentally that this smoothing efficiency limit is reached well before this theoretical limit. They therefore further improved the method by increasing the range covered by the three points. For example, if these are denoted by A_{i-1}, A_i and A_{i+1}, the spread can be increased to A_{i-2}, A_i and A_{i+2} and, more generally, to A_{i-j}, A_i and A_{i+j}. Clearly, the greater the value of j the more effective the smoothing, but the greater the loss in resolution. This point has been examined in detail by Maddams and Mead [11], whose results provide a clear guide to the permissible degree of smoothing.

12.2.3 The least-squares method

The effect of noise will be to lead to a scatter of individual data points about the best estimated line drawn through them. Hence, the determination of such a line, by a least-squares procedure, provides an effective method for the reduction of random noise, and this is the basis of the widely used method of Savitzky and Golay [12]. A set of n data points is fitted to a polynomial by a least-squares calculation. The value of n is usually considerably smaller than the total number of data points involved in defining the absorption band; hence, this set of points is moved forward sequentially by one data point until the whole band has been covered. This smoothing method is mathematically equivalent to convoluting the original data with a numerical function. Savitzky and Golay have provided values for these functions for polynomials of various degrees, and these make the computer programming a straightforward task. Steinier et al. [13] have subsequently noted errors in some of these numerical values and have supplied corrected tables.

12.3 Multi-component analysis

12.3.1 Principles

The absorbance of a solute at a particular wavelength, λ, is related to its concentration and the pathlength by Beer's law:

$$A_{1\lambda} = \epsilon_{1\lambda} bc$$

where ϵ_λ is the molar absorptivity at λ. If a second absorbing species is present, its absorption behaviour will be given by $A_{2\lambda} = \epsilon_{2\lambda} bc$ and, if there is no interaction between the two components, the total measured absorbance A_λ will be the sum of that of the two absorbances, $A_{1\lambda} + A_{2\lambda}$. More generally, for an n component system, there will be the relation:

$$A_\lambda = \epsilon_{1\lambda} bc_1 + \epsilon_{2\lambda} bc_2 + \ldots + \epsilon_{n\lambda} bc_n$$

Because there are n unknowns and only one equation, it is not possible to solve for $c_1, c_2, c_3 \ldots c_n$. However, if measurements are made at n wavelengths, the following set of simultaneous equations is obtained:

$$A_1 = \epsilon_{11} bc_1 + \epsilon_{21} bc_2 + \epsilon_{31} bc_3 + \ldots + \epsilon_{n1} bc_n$$
$$A_2 = \epsilon_{12} bc_1 + \epsilon_{22} bc_2 + \epsilon_{32} bc_3 + \ldots + \epsilon_{n2} bc_n$$
$$A_3 = \epsilon_{13} bc_1 + \epsilon_{23} bc_2 + \epsilon_{33} bc_3 + \ldots + \epsilon_{n3} bc_n$$
$$\ldots\ldots\ldots\ldots\ldots\ldots\ldots\ldots\ldots\ldots\ldots\ldots\ldots$$
$$A_n = \epsilon_{1n} bc_1 + \epsilon_{2n} bc_2 + \epsilon_{3n} bc_3 + \ldots + \epsilon_{nn} bc_n$$

In principle, therefore, $c_1, c_2, c_3 \ldots c_n$ may be determined. The degree to which this may be achieved in practice will now be considered.

12.3.2 The scope and limitations

One potential limitation is readily appreciated by considering the simplest multicomponent system, a two-component mixture, for which the two measured absorbances, A_1 and A_2, are given by the equations

$$A_1 = \epsilon_{11} bc_1 + \epsilon_{21} bc_2$$
$$A_2 = \epsilon_{12} bc_1 + \epsilon_{22} bc_2$$

Consider two extreme cases, the first in which component 2 does not absorb at λ_1 and component 1 does not absorb at λ_2. This gives the simple situation in which $A_1 = \epsilon_{11} bc_1$ and $A_2 = \epsilon_2 bc_2$, and the two components are, effectively, determined independently. The second extreme is when $\epsilon_{11} = \epsilon_{21}$ and $\epsilon_{12} = \epsilon_{22}$. There is then a total lack of specificity and only $c_1 + c_2$ may be determined. In choosing λ_1 and λ_2, it is therefore necessary to maximize the differences between the spectra of the two components, although it is unlikely that the ideal situation of zero absorption by one component will be achieved. It clearly becomes more difficult to maximize the specificity as the number of components increases and this factor, for a given error in the measurement of the various A and ϵ values, limits the precision that may be achieved.

It is common practice to choose λ_n, the characteristic wavelength for component n, in such a way that it lies at the maximum of a strong absorption band and is therefore insensitive to small errors in setting the monochromator. However, the way in which ϵ_{2n}, ϵ_{3n}, etc. vary with λ in the vicinity of λ_n must also be taken into account. Although these values may be appreciably smaller than ϵ_{1n}, if they vary rapidly with wavelength, because λ is located on the side of the absorption bands of other components, the fractional errors involved in their measurement may be considerably greater than for ϵ_{1n}

These points are well illustrated in the analysis of four component mixtures of ethylbenzene and o–, m– and p-xylene (Fig. 12.3). The choice of the characteristic (or key wavelength as it is sometimes called) is, in principle, easiest in the case of p-xylene. This has its longest wavelength band at 274.7 nm, a wavelength greater than those of the others, 271.0 nm for o-xylene and 272.7 nm for m-xylene, as may be seen from Fig. 12.3. At 274.7 nm, the absorptivities of ethylbenzene and o-xylene have fallen off considerably from the values at their maxima, and they are changing comparatively slowly with wavelength. However, 274.7 nm is on the side of the m-xylene band at 272.7 nm and ϵ for the latter is changing rather rapidly with λ. Nevertheless, this must be tolerated as 274.7 nm is the best overall choice for p-xylene.

Likewise, 272.7 nm proves to be the best choice for m-xylene. Although ϵ is changing more rapidly with λ for ethylbenzene and o-xylene than at 274.7 nm because 272.7 nm is closer to their respective maxima, fortuitously, this wavelength coincided with an absorption minimum in the p-xylene spectrum. The value of 271.0 nm proves to be the best overall choice for o-xylene, although it is clear that the specificity will be lower for

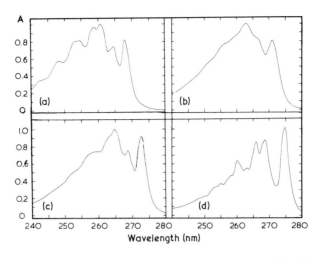

Fig. 12.3 Absorption spectra of four related compounds. (a) Ethylbenzene, $0.5129\,\text{g}\,\text{l}^{-1}$; (b) o-xylene, $0.3925\,\text{g}\,\text{l}^{-1}$; (c) m-xylene, $0.3947\,\text{g}\,\text{l}^{-1}$; (d) p-xylene, $0.1707\,\text{g}\,\text{l}^{-1}$.

the material than for the *m*- and *p*-isomers, and because ϵ_{max} is smaller. Ethylbenzene proves to be the most difficult, because of the overall weakness of its spectrum by comparison with those of the xylene isomers. There is little choice other than to use one of the two most obvious bands, at 261.5 and 268.6 nm. The results of measurements on synthetic mixtures bear out what may be surmised from the spectra of the four components, that the precision of the *p*-xylene determinations is the highest and that of the ethylbenzene the lowest.

It will be evident from this example that the feasibility of a particular multicomponent analysis is dependent both on the number of components and the specificity of their spectra. It is possible to form a good idea of what may be achieved in practice by calculations based on propagation of errors, but with any system of this type it is highly advisable to undertake background studies with synthetic mixtures in order to establish the errors in the estimation of the various components. In practice, the errors may be larger because of the presence of irrelevant absorption at a low level or a minor well-defined component. For example, iso-propylbenzene may be present in wider boiling xylene cuts and will interfere badly with the ethylbenzene determination.

Many three-component analyses have been reported and a few involving four components, but beyond this point, other analytical methods, particularly based on separation procedures, become markedly superior.

When a particular multicomponent analysis has been shown to be feasible and is in frequent use, it is possible to simplify the calculations involving the use of simultaneous equations by making use of the properties of matrices. The set of equations considered above may be written in matrix notation. For a given pathlength $b=1$, the equation becomes $\mathbf{A} = \mathbf{EC}$, where the vector $\mathbf{A}$ contains all the absorbances A_1 to A_n and likewise the concentration vector, $\mathbf{C}$, the concentrations of the components c_1 to c_n. The matrix of molar absorptivities $\epsilon_n \lambda_n$ is denoted by $\mathbf{E}$.

If this equation is premultiplied by $\mathbf{E}^{-1}$ it gives $\mathbf{C} = \mathbf{E}^{-1}\mathbf{A}$, so that the concentration values may be obtained directly. The inverse of this later matrix $\mathbf{E}^{-1}$, is easily calculated by standard methods of matrix algebra, as discussed in various standard texts, e.g. Aitken [14] and Bauman [15].

The analysis of an n component mixture by inverting an absorbance matrix of n rows and columns is mathematically equivalent to fitting the measured spectrum at n points, the characteristic wavelengths, with the spectra of the components. It is possible to improve the precision of the analysis considerably by fitting the spectrum as a whole using a least-squares criterion. In practice, this involves collecting absorbance values at close wavelength intervals over the whole of the spectral range of interest, preferably by digital means, and using a computer to obtain the least-squares fit. Surprisingly, although this method was described over 30 years ago [16], its merits have not been appreciated, but it is safe to predict that it will gain acceptance now that spectrometers with the requisite on-line data handling and computing facilities are coming into more general use.

There may be situations where deviations from Beer's law occur, for two reasons. Absorbance may not be a linear function of c, usually because of instrumental effects, e.g. the use of a slitwidth which is not small by comparison with the bandwidth or because of the use of very high absorbance values. In addition, there may be physical or chemical interactions between two or more components present in a mixture. Such interactions are more probable in IR spectroscopy, where concentrations of 10% are commonplace. Only in exceptional circumstances, e.g. highly hydrogen-bonding solvents and polar solutes, e.g. phenol in the presence of very low concentrations of hexamethyl phosphoramide, examined by Gerrard and Maddams [17], are interactions likely to occur at the low solute concentrations used in UV–VIS spectroscopy. When non-linearity occurs because of instrumental factors, it is possible to obtain reasonably quantitative analysis by the use of approximation procedures; a very useful account has been given by Bauman [18], and the calculations involved are trivial when handled by a computer.

12.4 Matrix rank analysis

The basis of the use of simultaneous equations for the analysis of multicomponent systems is that the number of components and their identities are known, but with unknown systems neither are available. Although analysis by other techniques will often provide the information, it may be possible to identify the components by UV–VIS spectroscopy by using spectral-stripping or curve-fitting techniques (see Section 12.5). Matrix rank analysis or 'factor analysis' can often provide this information more rapidly.

If absorbance measurements are made at a number of wavelengths on a series of solutions containing an unknown number of components whose concentrations vary with respect to each other, the results may be set out in matrix form. For example, for i such solutions and measurements at j wavelengths, the ij measurements constitute the rows and columns, respectively, of the matrix **A**, where

$$\mathbf{A} = \begin{bmatrix} A_{11} & A_{21} & A_{31} & \cdots & A_{n1} \\ A_{12} & A_{22} & A_{32} & \cdots & A_{n2} \\ A_{13} & A_{23} & A_{33} & \cdots & A_{n3} \\ \vdots & \vdots & \vdots & & \vdots \\ A_{1n} & A_{2n} & A_{3n} & \cdots & A_{nm} \end{bmatrix}$$

Such a matrix may be square, i.e. $i = j$, or rectangular, $i \neq j$. Each element of the matrix is the sum of the absorbances of the individual components in a particular solution at one wavelength. As noted in the previous section, the absorbance matrix is the product of two separate matrices, one of which **C** represents the concentrations of each component in the different mixtures and a second **E** which represents the molar absorptivities of each component at the various wavelengths.

If the number of wavelengths chosen and the number of solutions examined is greater than the anticipated number of components, at least one row or column of elements in the **C** and **E** matrices will consist wholly of zeros. However, this will not be apparent from the absorbance matrix, as is evident from the following simple example:

$$\begin{bmatrix} 7 & 17 & 26 \\ 10 & 14 & 24 \\ 13 & 11 & 22 \end{bmatrix} = \begin{bmatrix} 3 & 1 & 0 \\ 2 & 2 & 0 \\ 1 & 3 & 0 \end{bmatrix} \times \begin{bmatrix} 1 & 5 & 7 \\ 4 & 2 & 5 \\ 0 & 0 & 0 \end{bmatrix}$$
$$\quad\quad\mathbf{A} \quad\quad\quad\quad\quad\quad \mathbf{C} \quad\quad\quad\quad\quad\quad \mathbf{E}$$

The above example might well represent the results of measurements on a two-component system, measuring at three wavelengths on three different solutions. The number of components present is thus given by the rank, R, of the **A** matrix, which is defined as the order of the largest non-zero determinant computed from the matrix components, while all other determinants of order $R + 1$ are zero. This is a direct consequence of the fact that if the matrix is of rank R it requires the existence of R linearly independent absorbing components.

The value of R may be found by testing all of the determinants in the matrix of zero values. This is a time-consuming process, although it is readily amenable to computer calculation, and a further complication arises because measurement errors mean that a zero value is very rarely obtained. It is then necessary to use statistical tests to determine which are zero values. The method of Wallace and Katz [19] involves the setting up of a second comparison matrix whose elements S_{ij} are the estimated errors in the observed values of A_{ij}. The $\mathbf{A}_{ij}$ matrix is then transformed to an equivalent matrix, whose elements below the principal diagonal are all zero. The error matrix $\mathbf{S}_{ij}$ is also transformed during the reduction of $\mathbf{A}_{ij}$, by calculating new values for $\mathbf{A}_{ij}$ based on the propagation of errors during the transformation of $\mathbf{A}_{ij}$. The rank of $\mathbf{A}_{ij}$ is then found from the number element being given by the transformed $\mathbf{S}_{ij}$ matrix. Worked examples, which illustrate the method very clearly, are given by Wallace and Katz, who used the method to determine the number of species present in methyl red solutions of different pH values, and by Gerrard and Maddams [20], who examined phenol in solution in cyclohexane containing various concentrations of ethanol.

Although it should be possible to determine the rank of a matrix and the number of components present in a system by using the number of wavelengths equal to the rank, it is advisable with a completely unknown system to use a number of wavelengths significantly in excess of the likely number of components, for two reasons. It ensures that a more than adequate number of absorbance values is available and, as some wavelengths are more likely to be useful than others in terms of specificity for the individual components, it also increases the reliability on this count. The fact that the matrix is large is of no consequence because the calculations involved are readily handled by a minicomputer. However, it must be emphasized that matrix rank analysis will not give a greater degree of discrimination than is inherent in the measurements. If two components have very similar spetra, or a component is present in low concentration, there will be ambiguity. Two useful papers, by Ainsworth [21] and Burgess [22], deal with the application of matrix rank analysis to UV–VIS spectrometry, and Malinowski has published a detailed work on factor analysis [23].

12.5 Spectral stripping and related techniques

The very high performance of modern spectrometers, particularly with respect to sensitivity and signal-to-noise ratio, coupled with the availability of computers, off-or on-line, has led to the increased use of other mathematical methods for extracting information from the spectra of mixtures of components.

Spectral stripping provides a convenient introduction to these newer approaches, both as a method and because it provides a link between the manual and computer techniques. If the spectrum of a multicomponent mixture gives one or more clearly defined bands that can be assigned to specific compounds, these bands can be sequentially subtracted from the spectrum, giving a quantitative estimation, and also simplifying the residual spectrum so that it may be further interpreted. The subtraction can be done instrumentally or by computer. In the former case, a sample of the first of the identified components is placed in a variable pathlength cell in the reference beam of the spectrometer and the thickness is adjusted until the relevant band or bands in the mixture, sited in the sample beam, disappear. The process is then repeated with the second, and possibly other, components. The alternative approach of computer subtraction involves running the spectra of the identifiable components, storing them in digital form in the computer and using a scaling factor calculated by the computer to give exact subtraction from the composite spectrum. This approach offers greater flexibility and is less time consuming; it is a cogent example of the logical use of the computer in spectroscopy.

This technique is simply a multiple application of difference spectroscopy, but has limitations, some obvious and some a trap for the unwary. It will clearly work to the best advantage when the overall spectrum shows one or more well-defined bands that are not badly overlapped. This is a necessary but not sufficient condition, because the species responsible for these bands must not only be identified but must be available. As with the use of simultaneous equations, it is tacitly assumed that Beer's law is applicable. If there are interactions which lead to small shifts in the positions of the component bands in the mixture spectrum, the subtraction of the bands of the individual components can never be exact. The results will be akin to the shape of a first derivative spectrum. Fortunately, this effect is usually recognizable. During the subtraction process, the errors in both the original data and the inexactness of the subtraction become cumulative, and it is particularly important to avoid working with bands of very high or low intensity, as the spectrometric errors in the original spectrum will be greater. Finally, it is unwise to try to push the technique beyond reasonable limits. These cannot be defined uniquely as they vary from situation to situation. Perhaps the best working rule is that if the spectrum after a subtraction looks unusual or distorted, an error should be suspected.

Spectral stripping relies on the sequential subtraction of a series of identifiable peaks from a composite spectrum. It is possible to work in the opposite direction and match a multi-peaked spectrum by adding a series of computer-generated peaks until the resulting synthetic spectrum matches the experimental data. This process is known as curve fitting, and it is also subject to appreciable limitations, as has become increasingly recognized [24–27]. These limitations may be detailed by reference to Fig. 12.4 which shows a composite profile, together with the results obtained

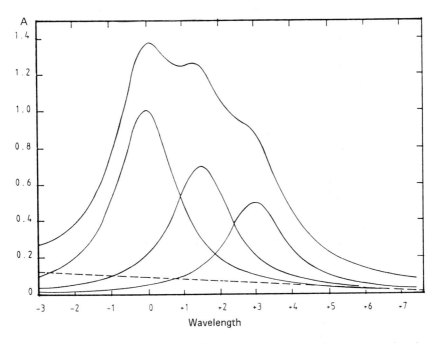

Fig. 12.4 Fitting a complex spectral shape by summation of components bands. The upper curve has been fitted by the sum of the three simple curves plus a linear sloping background.

by curve fitting. This has been done in terms of three peaks, a reasonable approach in view of the visual appearance of the composite profile. However, it is pertinent to ask if four or more peaks would have been more appropriate and, therefore, what criteria exist for establishing the number of peaks. It is also desirable to have approximate values for their locations on the wavelength axis, information that is usually obtained by visual inspection.

It is also necessary to establish two other sets of parameters for peaks, either by assumption or by an optimization procedure, as part of the calculations. These are the band shapes and half-widths. In the present instance, fitting has been done with peaks having the Lorentzian shape and of equal half-width but, in practice, it is necessary to justify a particular assumption. In Fig. 12.4 it has been necessary to include a baseline in addition to the three peaks, and this generally proves to be the case. The baseline has been taken as linear, but this may not be warranted. Hence, methods are required to define baselines and to deal with the general situation when they are not linear. Finally, a goodness of fit criterion is required to assess the overall success of the curve fitting and to show if any of the assumptions are seriously in error.

Most curve-fitting studies have related to the characterization of overlapping vibrational bands in IR and Raman spectra, and there is a definite change of emphasis when dealing with overlapping UV–VIS absorption bands. The band shape is different, the half-widths are much greater and the bands are more overlapped, but tend to be fewer in number. The primary spectral data have a markedly superior signal-to-noise ratio in the UV–VIS region. Some of these factors facilitate the use of curve fitting, some make it more difficult, and with others, e.g. the band shape, the problem is primarily that of justifying assumptions. These factors have been discussed in detail by Barker and Fox [28], who have done pioneering work in this field.

The greater the number of parameters to be optimized in the least squares process of curve fitting, the lower the probability that a unique solution will be forthcoming. It is therefore advantageous to fix, or at least to constrain, some of these parameters, and the number of component peaks is the one that offers the best possibilities. Visual inspection will often reveal almost hidden peaks that appear as slight shoulders, and the band sharpening achieved by the use of even-numbered derivatives is also valuable for estimating the number of component peaks. Recently, a promising new approach to band sharpening, termed Fourier self-deconvolution, has emerged. Any experimental peak can be expressed, mathematically, as a convolution of a lineshape function and a spectral function. If part or all of the former can be deconvoluted from the measured peak, a considerable sharpening, by as much as a factor of three, results. Hence, in the case of overlapped peaks, resolution of the components may be achieved to a degree that is valuable for peak finding prior to curve fitting.

On the grounds of mathematical convenience, the deconvolution is undertaken in the Fourier domain, and it is an added advantage that an increasing number of IR and NMR spectrometers have their primary output in this form. When the spectrometer output is in the conventional form of absorption intensity as a function of wavelength, it is collected in digital form and transformed by the use of the equation:

$$A(\lambda) = \int_{-\infty}^{\infty} I(x) \exp(i2\pi\lambda x) \, dx = F\{I(x)\}$$

The last parameter, $F\{I(x)\}$ is the Fourier transform of $A(\lambda)$. Such calculations present no problems with modern computers. The paper of Kauppinen et al. [29] should be consulted for fuller details. As yet, the technique has been used only with systems of overlapping vibrational bands, but it could prove useful in the case of composite band systems in UV–VIS spectra.

References

1. Mulder, E.J., Spruit, F.J. and Keuning, K.J. (1963) *Pharm. Weekblad.*, **98**, 745.
2. Morton, R.A. and Stubbs, A.L. (1947) *Biochem. J.*, **41**, 525.
3. Adamson, D.C.M., Elvidge, W.F., Gridgman, N.T., Hopkins, E.H., Stuckey, R.E. and Taylor, R.J. (1951) *Analyst*, **76**, 445.
4. Shaw, H.C. and Jefferies, J.P. (1953) *Analyst*, **78**, 519.
5. Morton, R.A. and Stubbs, A.L. (1946) *Analyst*, **71**, 348.
6. Ashton, G.C. and Tootill, J.P.R. (1956) *Analyst*, **81**, 232.
7. Ashton, G.C. and Tootill, J.P.R. (1956) *Analyst*, **81**, 225.
8. Glenn, A.L. (1963) *J. Pharm. Pharmacol. Suppl.*, **15**, 123T.
9. Clewes, B.N. (1979) *UV Spectrom. Grp Bull.*, **7**, 35.
10. Trott, G.W. and Beynon, J.H. (1979) *Int. J. Mass. Spectrom. Ion Phys.*, **31**, 37.
11. Maddams, W.F. and Mead, W.L. (1982) *Spectrochim. Acta*, **38A**, 437.
12. Savitzky, A. and Golay, M.J.E. (1964) *Anal. Chem.*, **36**, 1627.
13. Steinier, J., Termonia, Y. and Deltour, J. (1972) *Anal. Chem.*, **44**, 1906.
14. Aitken, A.C. (1944) *Determinants and Matrices*, 3rd edn. Oliver and Boyd, Edinburgh.
15. Bauman, R.P. (1962) *Absorption Spectroscopy, Appendix 1, Matrix Methods*. Wiley, New York.
16. Blackburn, J.A. (1965) *Anal. Chem.*, **37**, 1000.
17. Gerrard, D.L. and Maddams, W.F. (1978) *Spectrochim. Acta*, **34A**, 1219.
18. Bauman, R.P. (1962) *Absorption Spectroscopy*, pp. 413–419. Wiley, New York.
19. Wallace, R.M. and Katz, S.M. (1964) *J. Phys. Chem.*, **68**, 3890.
20. Gerrard, D.L. and Maddams, W.F. (1978) *Spectrochim. Acta*, **34A**, 1213.
21. Ainsworth, S. (1972) *Photoelec. Spectrom. Grp Bull.*, **20**, 611.
22. Burgess, C. (1979) *UV Spectrom, Grp Bull.*, **7**, 25.
23. Malinowski, E.R. (1991) *Factor Analysis in Chemistry*, 2nd edn. Wiley, New York.
24. Vandeginste, B.G.M. and DeGalan, L. (1975) *Anal. Chem.*, **47**, 2124.
25. Morrow, B.A. and Cody, I.A. (1973) *J. Phys. Chem.*, **77**, 1465.
26. Gans, P. and Gill, J.B. (1977) *Appl. Spectrosc.*, **31**, 451.
27. Maddams, W.F. (1980) *Appl. Spectrosc.*, **34**, 245.
28. Barker, B.E. and Fox, M.F. (1980) *Chem. Soc. Rev.*, **9**, 143.
29. Kauppinen, J.K., Moffatt, D.J., Mantsch, H.H. and Cameron, D.G. (1981) *Appl. Spectrosc.*, **35**, 271.

13 Special Techniques

The generally uninformative nature of the electronic envelope, coupled with the broad, overlapping nature of constituent absorption bands in mixtures, has led to the search for technical modifications, or spectroscopic 'tricks', for improving specificity and accuracy. Perhaps the earliest successful addition to the repertoire was Britton Chance's technique of dual-wavelength spectrometry [1] for measurements of turbid samples in biochemical analysis, later to be extensively applied by Shozo Shibata in analytical spectroscopy [2]. Shortly after Chance's pioneering work, two groups independently proposed the use of mathematical derivative functions in spectroscopy. In America, Giese and French [3] developed the earliest first–derivative device for examining the visible spectra of pigments in plant photochemistry. At the same time in the UK, Singleton and Collier patented the idea of using second- and higher-order derivative spectra [4], illustrated by their work in IR spectroscopy [5]. Difference spectroscopy, by contrast, has long been used for spectral correction purposes, represented in its simplest form by subtractive compensation of sample absorbance with that of the solvent. Recent improvements in instrumentation have led to the use of difference spectroscopy in the measurement of highly absorbing samples.

The impact of low-noise operational amplifiers and microcomputers on the practice of UV–VIS spectroscopy has been profound, not only as regards instrument control, but also with respect to the acquisition, storage and manipulation of digitized spectral data. A number of the spectroscopic 'tricks' discussed below have been made more accessible by virtue of digital techniques, which have undoubtedly transformed the practice of UV–visible spectroscopy.

13.1 Derivative spectroscopy

The derivative method is one of several generally applicable methods available for the transformation of spectral data. In this method, the transmittance, or more appropriately, the absorbance (A) of a sample in solution or in the gaseous state is differentiated with respect to wavelength, to generate the first-, second- or higher-order derivative:

$$A = f(\lambda)$$
$$dA/d\lambda = f'(\lambda)$$
$$d^2 A/d\lambda^2 = f''(\lambda) \ldots$$

Spectra transformed in this way often yield a more characteristic profile, where subtle changes of gradient and curvature are observed as distinctive bipolar features.

The first derivative of a UV–VIS absorption band, which represents the gradient at all points of the spectral envelope, has often been used to detect and locate 'hidden' peaks (since $dA/d\lambda = 0$ at peak maxima) and as a characteristic profile for identification [3, 6]. However, it turns out that the second derivative, which describes the 'curvature' of the original band, and the even higher derivatives are potentially more useful analytically [7].

The odd derivatives of a Gaussian band are observed as bipolar disperse functions dissimilar to the original band (Fig. 13.1). The even derivatives,

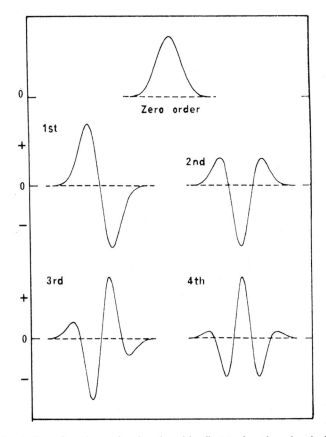

Fig. 13.1 A Gaussian absorption band and its first to fourth-order derivatives.

however, are seen as bipolar functions of alternating sign at the centroid, whose position coincides with that of the original band maximum. To this extent even derivative spectra bear a similarity to the original spectrum, although the presence of satellite peaks flanking the centroid adds a degree of complexity to the derivative profile. A key feature of derivative spectroscopy is the observation that, relative to the original spectral bandwidth in the so-called 'zero-order spectrum', the derivative centroid bandwidth decreases to 53%, 41% and 34% in the second, fourth and sixth orders, respectively [7]. This feature of the derivative process forms the basis of resolution for overlapping bands [8, 9], although it is clear that the band-sharpening process in Gaussian bands is less marked in the fourth and higher orders. Moreover, the increasingly complex satellite patterns detract from resolution enhancement in the higher derivative spectra.

An important property of the derivative process is that broad bands are suppressed relative to sharp bands, to an extent that increases with derivative order. This arises from the observation that the amplitude D_n of a Gaussian (or Lorentzian) band in the nth derivative is inversely related to the original bandwidth W, raised to the nth degree:

$$D_n \alpha W^{-n}$$

Thus, for two coincident bands of equal intensity, the nth derivative amplitude of the sharper band X is greater than that of the broader band Y by a factor which increases with derivative order:

$$\frac{D_{n,X}}{D_{n,Y}} = \left(\frac{W_Y}{W_X}\right)^n$$

This property leads to the selective rejection of broad, additive spectral interferences, and can lead to increased detection sensitivity in the second-and fourth-order derivatives [7, 9–11]. Other types of background interference, e.g. 'irrelevant absorption' or Rayleigh scattering, are also selectively suppressed in the derivative spectrum (Fig. 13.2).

The quantitative relationships which apply in derivative spectroscopy depend on whether transmittance or absorbance is the function of radiation intensity recorded. If transmittance is employed, the resulting expressions after differentiation become non-linear with concentration in the second and higher derivatives [7, 10]. If, however, absorbance (A) is employed, and if Beer's law is obeyed at any defined wavelength λ in the zero-order spectrum:

$$A = \epsilon bc \quad \text{at } \lambda$$

then

$$\frac{dA}{d\lambda} = \frac{d\epsilon}{d\lambda} bc \quad \text{at } \lambda$$

and

$$\frac{d^n A}{d\lambda^n} = \frac{d^n \epsilon}{d\lambda^n} bc \quad \text{at } \lambda$$

where ϵ is the molar absorptivity (M^{-1} cm^{-1}), b is the cell pathlength (cm) and c is the concentration (mol l^{-1}).

For quantitative work, the amplitude of a derivative peak can be measured graphically in various ways, as illustrated in Fig. 13.2 [7, 12]. Although the 'true' derivative amplitude is that measured with respect to the derivative zero, the most common practice is to record the amplitude with respect to satellite features of the spectrum (Fig. 13.2). This method affords an extra degree of suppression of matrix interference, since it is effectively a form of internal normalization of the derivative peak with respect to the overall derivative spectral envelope [7]. For highest accuracy, it is common practice to run standards in bracketing sequence with samples, subjecting both to the same experimental conditions. It should have been established that the graphical derivative measure adopted fulfils the conventional analytical criteria of linear response with concentration, regression through or close to the origin, independence from matrix or other interferences and optimum precision.

13.1.1 The production of derivative spectra

In general, methods for generating derivative spectra fall into two classes: optical methods which operate on the radiation beam itself; and electronic

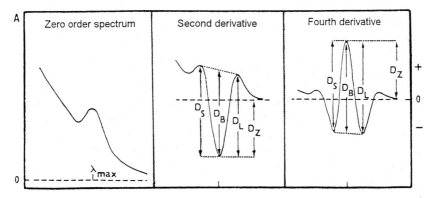

Fig. 13.2 A Gaussian absorption band superimposed on a light-scattering matrix ($A\alpha\lambda^{-4}$), and its second and fourth derivatives, illustrating the graphical measures available for quantitative assay. D_S and D_L are the amplitudes of the short- and long-wavelength satellites, D_B is the mid-point amplitude and D_Z is the derivative zero. These values are related to concentration through the derivative form of Beer's law.

or digital methods operating on the photometric detector output. The principal optical method is represented by the wavelength modulation technique, where the wavelength of incident radiation is rapidly modulated over a narrow wavelength range $\Delta\lambda$ by an electromechanical device (oscillating slits, grating or mirror). This method has found popularity in the construction of dedicated spectrometers for environmental monitoring, where the characteristic second derivative profile of polynuclear aromatic hydrocarbons, for example, aids rapid identification [13]. In another optical technique, the dual-wavelength spectrometer has been used to generate first derivative spectra, by scanning the spectrum with each monochromator separated by a small, constant interval $\Delta\lambda$, as discussed below [2].

The second and higher derivatives can, however, be more readily generated using low-noise analogue resistance–capacitance (RC) devices [10] or by digital techniques [12, 14, 15]. The electronic analogue device generates the required derivative as a function of time as the spectrum is scanned at constant speed ($d\lambda/dt$):

$$\frac{dA}{dt} = \frac{dA}{d\lambda}\frac{d\lambda}{dt}$$

$$\frac{d^2A}{dt^2} = \frac{d^2A}{d\lambda^2}\left[\frac{d\lambda}{dt}\right]^2$$

The 'true' value of the derivative is therefore related to the derivative amplitude observed under defined instrumental conditions, through an instrumental constant, K:

$$\frac{d^nA}{d\lambda^n} = K\frac{d^nA}{dt^n}$$

The analogue derivative amplitude is observed to vary with the following instrumental parameters: scan speed, slitwidth, RC device gain factor. Furthermore, the signal-to-noise ratio is found to degrade by approximately a factor of two in each successive derivative order [12], so that the conventional RC device is limited to generating fourth derivatives [7, 11]. This limit has been extended to eighth and ninth orders by the use of special electronic circuitry for enhanced performance in the higher derivatives [10].

An alternative method for generating derivative spectra is based on the microcomputer, employing one of a number of digital algorithms [9, 12, 14, 15] to produce smoothed spectral derivatives, either in real time or by post-run processing of the digitized spectrum. The digital approach is increasingly employed in contemporary spectrometers, due to the widespread adoption of microprocessors for instrument control and data handling, coupled with the addition of dedicated microcomputers for further data processing capability.

It should be noted that although transformation of a UV–VIS spectrum to its second or higher derivative often yields a more highly characteristic profile than the zero-order spectrum, the intrinsic information content of the data is not increased – indeed, some data, e.g. constant 'offset' factors, are lost. Rather, does the derivative method tend to emphasize subtle spectral features barely detectable by eye, by presenting them in a new and visually more accessible way (Fig. 13.3). Moreover, the derivative method is generally applicable in analytical chemistry and can be used equally for resolution enhancement of chromatographic [7] or densitometric [16] data.

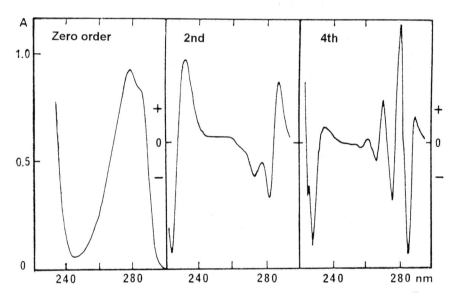

Fig. 13.3 The absorption spectrum (zero order) and second and fourth derivatives of ethinyl oestradiol in methanol ($140\,\mu g\,ml^{-1}$).

Derivative UV–VIS spectroscopy has found significant application in the environmental [13], pharmaceutical [10, 17–20], clinical [21], forensic [22], biomedical [10, 23] and industrial areas [11]. The method can be combined with difference or dual-wavelength spectroscopy, to give enhanced discrimination against matrix interference, as discussed below.

13.2 Difference spectroscopy

In conventional spectrometry, the absorbance of a liquid sample in a suitable cell is compensated by automatic (or manual) subtraction of solvent (A_s) and window (A_w) absorption at a defined wavelength, using a matched cell filled with the appropriate solvent. In this sense, all absorbance measurements in UV–VIS spectrometry are made by means of difference spectroscopy, where the difference in absorbance, ΔA, is equivalent to the absorbance of the analyte alone, A_X:

$$\Delta A = A_X + A_s + A_w - (A_s + A_w) = A_X \quad \text{at } \lambda$$

The difference spectroscopic method is sometimes described as 'differential spectroscopy', an ambiguous term not to be recommended in view of the increasing exploitation of derivative spectroscopy (cf. Section 13.1).

In general, difference spectroscopy involves the measurement of an absorbance difference between a liquid sample and a reference solution. The latter may consist of the sample solution in physically or chemically changed form, or it may comprise a solution equivalent in composition to the liquid matrix in which the sample is located. Two of the major types of application are represented by (a) matrix compensation methods, and (b) high-precision measurement of high-absorbance values. For convenience, each application type will be considered separately.

13.2.1 Matrix compensation methods

These methods generally exploit a change in the chemical or physical properties of the analyte X, which permits its selective detection in the presence of interfering matrix M. The assumption is made that the spectrum of the analyte can be changed without affecting the matrix spectrum. If, for example, the analyte is pH sensitive at a defined wavelength, with molar absorptivity values of ϵ_X at pH_1 and ϵ'_X at pH_2, and if the effective molar absorptivity of the matrix, ϵ_M, remains constant at both values of pH, then assuming that the law of additivity of absorbances applies, the total absorbance A_T observed at pH_1 and A'_T observed at pH_2 becomes:

$$A_T = \epsilon_X bc_X + \epsilon_M bc_M$$
$$A'_T = \epsilon'_X bc_X + \epsilon_M bc_M$$

The difference absorbance, ΔA (measured with the less absorbing solution in the reference cell), becomes:

$$\Delta A = |A_T - A'_T| = |\epsilon_X - \epsilon'_X|bc_X \quad \text{and} \quad \Delta A = \Delta\epsilon_X bc_X$$

Values for $\Delta\epsilon_X$ can be established by prior standardization, or in situ for every assay. If Beer's law holds at the analytical wavelength employed, then the concentration of analyte in a test can be readily found by simple proportion:

$$\Delta A_\text{test}/\Delta A_\text{standard} = C_{X(\text{test})}/C_{X(\text{standard})}$$

Many suitable methods for physical and chemical modification of the analyte absorbance have been reported. As an example, the classical method described by Görög for the determination of Δ^4- and $\Delta^{1,4}$-3-ketosteroids in pharmaceutical applications, illustrates the general approach involved [24]. A methanolic solution of a sample containing prednisolone is reduced by sodium borohydride to yield the spectrally inactive 3-hydroxy derivative. The residual absorption (A'_P) at about 240 nm is equivalent to the absorption of the unchanged matrix (A_M), so that when the reduced solution is placed in the reference cell, the untreated prednisolone sample absorbance (A_P) is compensated by difference for the matrix interference:

$$A_T = A_P + A_M$$
$$A'_T = A'_P + A_M$$
$$\Delta A = A_T - A'_T = A_P \quad \text{since } A'_P = 0$$

This elegant procedure has been modified by Chafetz et al. [25], who substituted lithium borohydride in tetrahydrofuran for more efficient reduction of $\Delta^{1,4}$-3-ketosteroids. In order to establish that the matrix spectrum is unchanged by the reducing agent, a plot of log (ΔA) as a function of wavelength can be compared with the logarithmic spectrum of the pure component, these curves being superimposable if the matrix interference has been eliminated. This follows from the observation that the shape of a logarithmic spectrum is defined only by the molar absorptivity as a function of wavelength, the concentration term being expressed as a vertical displacement along the log A axis:

$$\log_{10}(\Delta A) = \log_{10}(\Delta \epsilon) + \log_{10} b + \log_{10} c \quad \text{at } \lambda$$

Indeed, small differences between the two curves could be accentuated by transforming each log (ΔA) spectrum to its second derivative [7].

In analytical biochemistry, where difference spectroscopy is widely used for investigations on the effect of solvent, heat and other perturbing factors, transformation of the difference spectrum to its first or second derivative can be successfully achieved, and has permitted the examination of tyrosyl residues resolved in cytochrome P450 [26] and phenylalanine residues resolved in protein spectra, respectively [27]. Second derivative-difference spectroscopy, as the technique is described, affords a further method for rejecting matrix interference, while simultaneously sharpening any fine structural features and improving their resolution from adjacent bands.

In its simplest form, difference spectroscopy is practised in industrial quality control, in those cases where the sample matrix is well-defined. An appropriate dilution of the matrix is placed in the reference cell, on the assumption that the matrix composition can be accurately replicated and is not subject to variation during the industrial process. The difference absorbance is, however, susceptible to systematic error introduced by any uncertainty in the concentration of the matrix in the sample to be assayed. This error increases in proportion to the ratio of the molar absorptivities of matrix to analyte.

13.2.2 High-precision measurements

High-precision measurements of highly absorbing solutions can be achieved by using as reference a standard solution slightly lower in concentration than the test sample, coupled with expansion of the transmittance or absorbance scale and a wider slitwidth to increase energy throughput [28]. In principle, if the analyte obeys Beer's law up to the absorbance value of test (A_t) and standard (A_s), then the difference absorbance, ΔA, is related to analyte concentration in test (C_t) and standard (C_s) as follows:

$$\Delta A = A_t - A_s = \epsilon b (C_t - C_s) = \epsilon b \, \Delta c$$

In practice, stray-light, instrument noise and limits of photometric linearity require that a calibration curve of ΔA versus Δc be established. Various modifications of this technique have been proposed for high-sensitivity trace analysis [29].

13.3 Dual-wavelength spectroscopy

In this method, two independently monochromated beams of radiation at λ_1 and λ_2 are time-shared through a single sample cell. The difference in absorbance, ΔA, is recorded at the two analytical wavelengths selected, with the aim of reducing or eliminating matrix interference, or for the determination of analytes in a multi-component mixture:

$$\Delta A = A^{\lambda_1} - A^{\lambda_2}$$

The sample cell is usually positioned close to the detector so that turbidity or scattered radiation can be effectively compensated [30].

Dual-wavelength spectroscopy has been shown to permit accurate measurement of small absorbance differences, both in highly absorptive solutions and in very weakly absorbing systems [2, 30]. The 'reference' wavelength, λ_2, confers great flexibility on the method. In the presence of Rayleigh scattering, for example, the reference wavelength is set to a point on the spectral profile, which is equiabsorptive with the anticipated scattering contribution, $A_R^{\lambda_1}$, at the 'analytical' wavelength, λ_1:

$$A_T^{\lambda_1} = A_X^{\lambda_1} + A_R^{\lambda_1} \quad \text{at } \lambda_1$$
$$A_T^{\lambda_2} = A_R^{\lambda_1}$$
$$\Delta A = A_T^{\lambda_1} - A_T^{\lambda_2} = A_X^{\lambda_1}$$

where the subscripts T, X and R refer to the total absorbance, and the analyte and Rayleigh scattering contributions, respectively (Fig. 13.4).

A further example of matrix correction is the case where the reference wavelength is set at a point λ_2 in the spectrum where the matrix absorption, $A_M^{\lambda_1}$, behaves similarly to the interference, $A_M^{\lambda_2}$, observed at the analytical wavelength, λ_1:

$$A_T^{\lambda_1} = A_X^{\lambda_1} + A_M^{\lambda_1}$$
$$A_M^{\lambda_2} = A_M^{\lambda_1}$$
$$\Delta A = A_T^{\lambda_1} - A_M^{\lambda_2} = A_X^{\lambda_1}$$

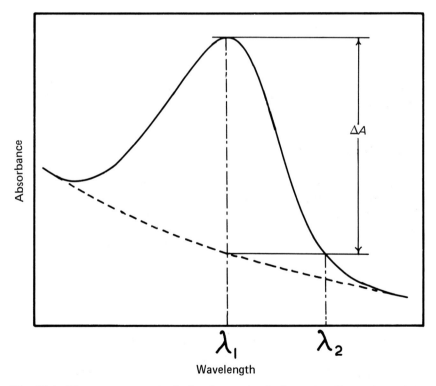

Fig. 13.4 The measurement of absorbance in dual-wavelength spectrometry. Redrawn from Shibata [2].

A variant of this method involves scanning the spectrum in difference mode, keeping the reference wavelength constant at a spectral position which constitutes a reference for the system [2]. Such 'relative' spectra correspond to a form of internal normalization.

In cases where two defined components X, Y overlap, the spectrum of Y may be deconvoluted or suppressed by selecting, for λ_1 and λ_2, wavelengths which correspond to equi-absorptive points on each side of a peak maximum in the unwanted component. Under these conditions, the difference absorbance for Y must be zero:

$$\Delta A = A_Y^{\lambda_1} - A_Y^{\lambda_2} = 0$$

Thus, the difference absorbance for component X will be independent of Y (Fig. 13.5):

$$A_T^{\lambda_1} = A_X^{\lambda_1} + A_Y^{\lambda_1} \quad \text{at } \lambda_1$$
$$A_T^{\lambda_2} = A_X^{\lambda_2} + A_Y^{\lambda_2} \quad \text{at } \lambda_2$$

and

$$\Delta A = A_X^{\lambda_1} - A_X^{\lambda_2}$$
$$= \Delta \epsilon_X \, bc.$$

Dual-wavelength spectroscopy can, in this example, be seen as a special case of difference spectroscopy. Various modifications of this approach to spectral suppression have been proposed, using analogue absorbance multipliers [2]. A method proposed for examining complex formation is to

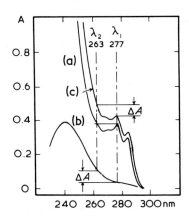

Fig. 13.5 The absorption spectra of (a) isophthalic acid (10% w/v), (b) terephthalic acid (0.5% w/v), and (c) a mixture containing both compounds in the same concentrations. Redrawn from Porro [30].

use an isosbestic point as a reference wavelength, in order to increase measurement accuracy and precision [2].

In dual-wavelength spectroscopy, the option to use each wavelength independently is available. This feature has been used to monitor two reactants at their respective optimal absorption maxima [2]. However, the more common mode is to measure the difference absorbance at two wavelengths, one set to the analyte and the other to a reagent which is consumed during the reaction. This method permits the sensitivity to be readily increased [2].

If the two monochromators are set a small spectral interval apart ($\Delta\lambda \approx 1-4$ nm), the first derivative spectrum, $dA/d\lambda$, can be readily obtained by scanning the monochromators in tandem. Moreover, the first derivative of absorbance at fixed wavelength can also be monitored for reaction kinetics. However, the higher derivatives are not accessible by this method, as discussed in Section 13.1.

Although dual-wavelength spectroscopy offers a number of intriguing measurement possibilities in UV–VIS spectroscopy, the cost of instrumentation has precluded its widespread use. The method appears to have found wide application for high sensitivity and high accuracy measurements in complex biological systems, particularly in plant photochemistry.

13.3.1 *Future perspectives*

The advent of rapid scanning spectrometers based on the linear photodiode array or the TV vidicon tube, coupled with powerful microcomputers for data storage, manipulation and presentation, opens up new perspectives in analytical spectrometry. In addition to the capability to generate second-and higher-derivative spectra, difference spectra and log spectra, these devices can be used for matrix least-squares deconvolution of defined, multicomponent mixtures using either zero-order or derivative spectra. The establishment of digital archives of spectra, coupled with rapid archive retrieval routines based on derivative spectra, will open up another phase of development in UV–IS spectroscopy. Rapid scanning detectors have also shown significant potential for enhanced detection capability in HPLC, where transformation of captured spectra to their second or higher derivative offers a more highly characteristic profile for identification [31].

13.4 Densitometry

Accessories for the measurement of the optical density of solid samples, e.g. photographic film, TLC plates or electrophoresis gels, are available for some spectrometers. As with all accessories, the spectrometer-plus-attachment is less satisfactory than a custom-built densitometer, particularly because the sample must be mounted vertically, it is difficult to align the beam through the region of interest, and a densitometer will have microscope optics allowing a higher beam intensity and smaller spot size to be used. On the other hand, the spectrometer has some advantages over densitometers that only operate with a measuring beam of white light, for the optimum wavelength for a particular measurement can be selected in order to minimize the effects of light-scattering, and UV-absorbing compounds can be measured directly without the necessity of staining them. The spectrometer can also measure the complete absorption spectrum of a particular spot, though the high background absorption and scatter will probably mean that the spectrum is of poor quality.

In general, the sample is mounted vertically in the measuring beam of the instrument, and the accessory will have some means of moving the sample horizontally so that it is scanned along its length. This motion requires a spectrometer with a large sample compartment and widely separated measuring beams, and consequently not all instruments can accommodate such an accessory.

Two common applications of densitometry in chemical and biochemical laboratories are the measurement of spots on a TLC plate, which may be coloured or stained, and the measurement of bands in electrophoresis gels. The latter may be in the form of a 'slab' – a sheet 1–2 mm thick – or a 'tube', which is a gel cylinder about 6 mm in diameter. These are usually stained with a blue dye, although in some cases direct measurements of the absorption of the bands can be made. The task is to measure the optical density of the separated spots or bands, and hence the concentration of the components of a mixture, and to identify these components by the location of the spot on the sample. Thus, the densitometer scan will consist of a plot of optical density or transmission of the sample against distance from some reference point. These samples are 'difficult' in spectrometric terms, since the gels or TLC plates are generally opaque and scatter light badly. It is therefore an advantage to use light of as long a wavelength as possible and the blue dyes used in gel electrophoresis with $\lambda_{max} > 600$ nm are ideal. In some cases, a better approach is to photograph the gel by either transmitted or reflected light, and scan the resulting negative. The operation is simple if a Polaroid camera with negative film is used. The loss of resolution caused by this extra process can be compensated by the increased contrast of the negative, and the negative is far easier to mount in the spectrometer. This technique is particularly useful if fluorescent bands are to be measured, the gel being photographed with an appropriate exciting wavelength while the camera is fitted with a filter to block the exciting light.

References

1. Chance, B. (1954) *Rev. Sci. Instrum.*, **22**, 634.
2. Shibata, S. (1976) *Angew. Chem. Int. Ed. Engl.*, **15**, 673.
3. Giese, A.T. and French, C.S. (1955) *Appl. Spectrosc.*, **9**, 78.
4. Singleton, F. and Collier, G.L. (1956) *Brit. Pat.*, **760**, 729.
5. Collier, G.L. and Singleton, F. (1956) *J. Appl. Chem.*, **6**, 495.
6. Olson, E.C. and Alway, C.D. (1960) *Anal. Chem.*, **32**, 370.
7. Fell, A.F. (1980) *UV Spectrom. Grp Bull.*, **8**, (P 1), 5.
8. Morrey, J.R. (1968) *Anal. Chem.*, **40**, 905.
9. Butler, W.L. and Hopkins, D.W. (1970) *Photochem. Photobiol.*, **12**, 439, 451.
10. Talsky, G., Mayring, L. and Kreuzer, H. (1978) *Angew. Chem. Int. Ed.*, **17**, 785.
11. Ishii, H. and Satoh, K. (1982) *Fres. Z. Anal. Chem.*, **312**, 114.
12. O'Haver, T.C. and Green, G.L. (1976) *Anal. Chem.*, **48**, 312.
13. Hawthorne, A.R. (1980) *Am. Ind. Hyg. Assoc. J.*, **41**, 915.
14. Savitzky, A. and Golay, M.J.E. (1964) *Anal. Chem.*, **36**, 1627.
15. Steiner, J., Termonia, Y. and Deltour, J. (1972) *Anal. Chem.*, **44**, 1906.
16. Traveset, J., Such, V., Gonzalo, R. and Gelpí, E. (1981) *J. Chromatogr.*, **204**, 51.
17. Traveset, J., Such, V., Gonzalo, R. and Gelpí, E. (1980) *J. Pharm. Sci.*, **69**, 629.
18. Fell, A.F. (1978) *Proc. Anal. Div. Chem. Soc.*, **15**, 260.
19. Fell, A.F. (1982) *Proc. Symp. Analysis of Steroids*, Eger, Hungary (ed. S. Görög), p. 459. *Elsevier Science, Amsterdam.*
20. Davidson, A.G. and Elsheikh, H. (1982) *Analyst*, **107**, 879.
21. O'Haver, T.C. (1979) *Clin. Chem.*, **25**, 1548.
22. Gill, R., Bal, T.S. and Moffat, A.C. (1982) *J. Forensic Sci. Soc.*, **22**, 165.
23. Fell, A.F. (1983) *Trends Anal. Chem.*, **2**, 63.
24. Görög, S. (1968) *J. Pharm. Sci.*, **57**, 1737.
25. Chafetz, L., Tsilifonis, D.C. and Riedl, J.M. (1972) *J. Pharm. Sci.*, **61**, 148.
26. Ruckpaul, K., Rein, H., Ballou, D.P. and Coon, M.J. (1980) *Biochim. Biophys. Acta*, **626**, 41.
27. Ichikawa, T. and Terada, H. (1979) *Biochim, Biophys. Acta*, **580**, 120.
28. Willard, H.H., Merritt, L.L. and Dean, J.A. (1974) *Instrumental Methods of Analysis*, p. 94. *Van Nostrand, New York.*
29. Donbrow, M. (1967), *Instrumental Methods in Analytical Chemistry, Vol. II. Optical Methods*, p. 130 *Pitman, London.*
30. Porro, T. (1972) *Anal. Chem.*, **44**, 93A.
31. Fell, A.F., Scott, H.P., Gill, R. and Moffat, A.C. (1972) *Chromatographia*, **16**, 69.

14 Automated Sample Handling

14.1 Introduction

Laboratories often meet large, unexpected increases in the demand for a particular analytical determination. In many of these instances, it has only been possible to respond to these demands within the required time by recourse to automated systems for sample handling. A close examination of existing practices may reveal advantages to be gained by the automation of sample-handling, thus releasing trained staff for more demanding tasks. Some laboratories have the resources to develop and construct their own dedicated systems, but most laboratories depend on the purchase of suitable equipment from the fairly wide range of modules which are available. Adaptations to commercial equipment can be made to render it more suited to specific requirements.

Automated sample handling systems can be divided into the following broad categories:

(a) Discrete systems.
(b) Centrifugal systems.
(c) Continuous-flow systems with segmentation.
(d) Continuous-flow systems without segmentation.

This chapter contains brief descriptions of the first two types and a more detailed account of continuous flow systems.

Discrete systems are those which closely reproduce the classical operations of pipetting, reagent addition, treatment and measurement by electro-mechanical methods.

Centrifugal systems are those in which samples and reagents are brought together and transferred to an optical cell on a rapidly rotating turntable. Absorbance measurements are made whilst the cells are rotating, and resolution of the individual signals is carried out electronically.

Continuous-flow (CF) systems are those through which a continuous flow of liquid is pumped and into which the sample solutions are successively inserted. Following subsequent addition of reagents and diluents and a variety of treatments, the stream passes through the flow cell of an appropriate detector.

CF automated analysis was introduced by Skeggs in 1957 [1] and commercially exploited at first by Technicon as the Auto Analyzer (R). Since its introduction, the Auto Analyzer has undergone a number of improvements leading to standard laboratory systems which can achieve sampling-rates of up to 60 per hour and advanced multi-analysis clinical systems which can perform up to 20 simultaneous determinations at rates of up to 150 samples per hour. These developments are all based on air-segmented systems and are due to the increased understanding of the scientific principles involved [2–10].

During the later period of development of air-segmented systems, there has been a fairly rapid growth in the development of non-segmented CF systems, under the general description of flow injection analysis (FIA). In this technique, samples are introduced into the flowing stream by means of the injection loop of a liquid sampling valve. Thus, a fixed volume of the sample passes through the system as a discrete slug as opposed to being distributed over a series of liquid segments corresponding to a given time of sampling, as in the segmented systems. This type of system will be described in more detail later.

14.2 Air-segmented continuous-flow systems

14.2.1 Introduction

Air-segmented CF analysers are the most widely used and versatile systems. Hundreds of papers have been published on the theory and applications of this technique, but there may still be a number of laboratories where the advantages have not been fully appreciated. The main aim of any automated system must be to maximize output. Many determinations require strict adherence to a rigorous timing schedule, and while humans may be fallible in this respect, an auto-analyser excels. Some determinations entail the handling of reagents which, even if not hazardous, may constitute a nuisance in the laboratory, and the use of a closed system with appropriate waste-disposal can often be beneficial. CF systems certainly act in the interests of health and safety, for they are capable of handling a range of hazardous liquids. However, any leakage or blockage of the system may lead to a burst, which could be dangerous unless appropriate safety measures are taken.

A schematic layout of a typical laboratory system is shown in Fig. 14.1 and with more detail in Fig. 14.2.

The sampler has to be capable of taking up solutions from successive sample cups for a pre-determined period of time, followed by an intermediate sampling of appropriate wash liquor, also for a predetermined length of time. These times must be accurately reproducible, and the time of passage of the probe between sample and wash must be minimized so as

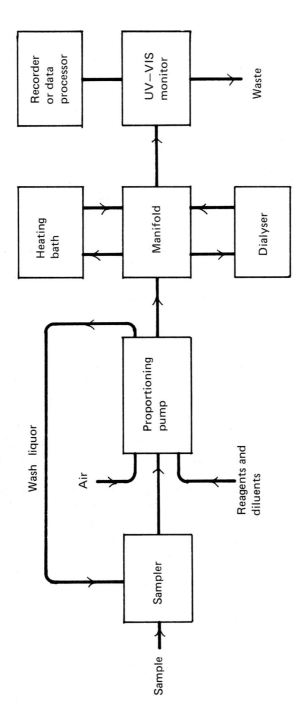

Fig. 14.1 Schematic diagram of an air-segmented continuous-flow analysis system.

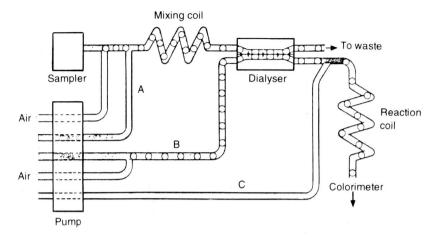

Fig. 14.2 Schematic diagram of an air-segmented continuous-flow analysis system using sample separation by dialysis. From Snyder *et al.* [10].

to limit the amount of air drawn into the system during this movement. A small air bubble does, however, help to sweep the sample line. Most samplers have a rotary sample plate, with peripheral holes to accommodate the plastic sample cups. Some have two concentric rings of holes and twin probes as a means of carrying out two simultaneous determinations. It is worth obtaining the type of plate with holes to take either 2 or 5 ml cups, since certain applications call for high sampling rates when the larger cups become essential.

Some samplers are timed by means of rotating cams and microswitches, while others use electronic timing. Cams provide a very reliable and robust method, but lack the total flexibility of electronic systems. It is possible to use external electronic timers to replace the use of cams. A correct choice of sampling and wash times is vital for optimal operation of the system, details of the method of deriving minimum sampling times will be found in the references. Wash times should normally be just sufficient to give adequate peak separation. The chosen times can be validated by running three 'low' samples, followed by three 'high' samples and a further three 'low' samples. Any unacceptable carry-over will be evidenced by the first 'high' peak being pulled down by the preceding 'low' and vice versa.

Some samples may produce a sediment whilst awaiting their turn to be sampled. Agitator attachments are available which will enable the sample to be homogenized during sampling. The sampling of viscous liquids can give rise to carry-over problems in the non-segmented sample line. One sampler exists which can be made to operate in a 'pecking' mode so that air-segmentation is introduced at the sample probe. Conventional samples are totally unsuitable for unstable samples which may have to reside on the turntable for an hour or so awaiting their turn. Newer types of auto-sampler are available which prepare the sample solution just prior to the actual sampling. A sampler is available for the automatic sample preparation from solids, e.g. tablets and capsules.

14.2.2 Pump systems

The heart of any system is the multi-channel peristaltic pump. Its function is to meter the sample and reagents or diluents, and to generate the essential regular, correct bubble pattern. The ideal bubble is slightly longer than its diameter and the optimum bubble frequency is determined by the type of manifold [5]. The Technicon Pump III uses an 'air-bar', which is designed to inject the air at an optimum rate for a system based on 2.0 mm internal diameter (i.d.) components. Other systems rely on a correct choice of relative liquid and air pumping rates to achieve the best bubble patterns.

Pumps are available which will accommodate as many as 28 pump tubes in a single layer. Earlier designs of pumps have their tubes mounted in two layers; the actual capacity of the pump, rather than the rated capacity, is often dependent on the particular combination of pump tubes required. The pumping rate is dependent upon the actual diameter of the pump tube. The sample, reagent and diluent volumes used in the related manual method must be converted into corresponding flow rates. For this purpose, colour-coded pump tubes are available covering a wide range of pumping rates. They are available in Tygon (for normal liquids), Solvaflex (for certain solvents) and Acidflex (for strong acids). Silicone–rubber pump tubes are a more recent addition; they are relatively expensive but can be used with liquids which are not suited to the other materials. 'Flow-rated' pump tubes have a closer tolerance on the pumping rate than the standard range, but are not available in silicone–rubber. The extra cost of flow-rated tubes is often worthwhile for more exacting analyses. Pumping difficulties may be experienced if attempts are made to utilize the smallest of the available range of pump tubes. The exact 'cut-off' size will depend on the choice of pump.

There is one problem in the unattended shutdown of a pump, which is the inability to release the roller pressure automatically. If the pump is stopped for any length of time without relieving the pressure, the tubes become permanently damaged. There are two ways to avoid the wastage of expensive reagents without endangering the pump tubes on shutdown: one is the use of special valves which may be operated manually or automatically to switch the pumping from reagents to wash liquid. The other method is to switch the pump into a stand-by mode in which it continues to operate intermittently. This latter facility is useful when a system may be needed again at short notice.

14.2.3 The manifold

This is the assembly of glass fittings, mixing coils and interconnecting tubing which forms the arterial liquid-handling system. Two ranges of components for standard laboratory systems are available, termed System I and System II. Many laboratories have retained their early System I equipment, and replacement modules and fittings can still be obtained. The i.d. of these components is about 2.3 mm. The System II components are generally smaller, with an i.d. of 2.0 mm. Lower pumping rates are used with a saving in reagent consumption, while better wash characteristics are obtained thus allowing an increased throughput of samples.

System I manifolds are normally constructed on a tray mounted on the pump. System II manifolds can be constructed in, and on, a special manifold module or 'cartridge'. Usually it is not possible to devote an auto-analyser to a single determination and regular changes of function are necessary. Although, with practice, System I manifolds can be changed over fairly conveniently, the storage of spare manifolds can be a problem. The System II cartridges are much more convenient, in this respect, and they can also include the smaller System II heating baths and dialysis units. It is advisable to keep the manifold and pump tubes assembled, so spare end-blocks for the pump are useful; the tubes can be taped to the end-blocks to prevent them getting tangled during storage.

The standard transmission tubing is 1.6 mm i.d. Tygon. Tubing in Solvaflex, Acidflex and silicone–rubber is also available. Polythene and glass are thought to offer better wash characteristics, but are obviously more difficult to work with from the point of view of bending and joining. Various 'bends' are available in glass transmission tubing, which ease the problem of constructing all-glass manifolds. Joints in the flexible tubing are made using the range of small connectors (nipples) which are available in plastic, glass and stainless steel–platinum. With Tygon and Solvaflex, sleeved butt-joints can be made using cyclohexanone to weld tubing and sleeving together.

14.2.4 Heating

Owing to the precise reproducibility of these systems, it is generally held to be unnecessary to heat the reaction mixture for as long as it takes to achieve 100% completion of the reaction. This is generally true, but a word of caution is necessary. In some instances, unless the reference solution is identical in composition to the samples, the rates of reaction can differ and erroneous results may be obtained. Such problems can be minimized by allowing the reaction to go to completion.

System I heating baths are large separate modules, accommodating coils of up to 25 m in length. System II heating baths are small, sealed units which can be accommodated in the manifold cartridge unit. However,

System II heating baths are often too small to give long delay times and the larger type of bath may have to be utilized. The size of coil required for a given residence time can be calculated from the total flow rate and its i.d. It is usual to categorize the smaller System II coils by volume and the larger coils by length.

14.2.5 Dialysis

Dialysis has been used as a convenient means of separating the test substance from irrelevant high-molecular weight material. As with heating baths, the larger System I modules are still available. System II units are much smaller and can form a part of the manifold. A variety of membranes is available to suit particular applications. Figure 14.2 shows a typical system. The proportion of the analyte of interest transferred across the membrane is quite low since the dialysis unit cannot be used in a counter-current mode and since some of the dialysers are quite small. However, this can be an advantage as it may avoid the need for a dilution stage. The flow rate of donor and recipient streams should be matched as closely as possible, which may call for the use of a slightly higher pumping rate for the recipient stream due to back pressure from the rest of the system.

14.2.6 Detectors

CF analysis was initially designed for colorimetric detection, but any method of detection can be used which will give a measurable signal and for which a suitable flow-cell can be constructed. Various colorimeters are available mostly based on interference filters. One colorimeter offers at least four channels, operating from a single light-source. For complete versatility, the use of a double-beam UV–VIS detector is preferable to a colorimeter. Any stable spectrometer may be used and a range of flow cells is available for most instruments.

It is usual to 'debubble' the stream, before drawing a portion through the cell, although debubbling has an adverse effect on sampling rates. The technique of 'bubble-gating' is now introduced, whereby the air is permitted to pass through the cell and the effects filtered out electronically [11, 12].

14.2.7 Data processing

Owing to the prolific output of data from CF systems, the time devoted to manual calculations can compare very unfavourably with the analysis time, and electronic data processing becomes imperative. Microprocessor systems are now available which are capable of controlling the system and monitoring its performance in addition to the generation of results. Under the heading of data processing, one could include the technique of 'curve regeneration'. It is normal in CF analysis to sample for a sufficient time to allow the response to effectively reach the maximum value, the so-called 'steady state' condition. This can limit the throughput of a system. The curve regenerator [13, 14] allows the use of reduced sampling times by continuously calculating the height of the steady-state value from the rate of increase in the response. Increases of up to 50% in throughput can be obtained, though for precise work, such systems must be carefully validated.

14.3 Flow injection analysis

Until relatively recently, it was assumed that segmentation was an essential feature of CF analysis, but FIA is a viable alternative. In this technique, the sample is inserted as a slug into a flowing stream of suitable carrier liquid containing the reagent by means of a rotary valve with an injection loop. A schematic system is shown in Fig. 14.3. FIA was developed independently by Stewart *et al*. in the USA, and Ruzicka and Hansen in Denmark. The former chose tubing with an i.d. of less than 0.5 mm. Since pumping pressures are in the region of 100–500 psi, an HPLC type of pump is required. The latter group use 0.5–1.0 mm tubing and a peristaltic pump is suitable. Commercial systems are available, based on both principles. Betteridge [15] has written a comprehensive report on this technique.

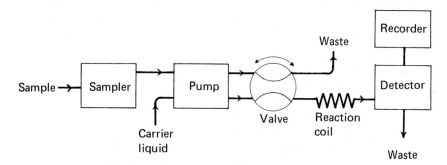

Fig. 14.3 Schematic diagram of a flow injection analysis system.

The advantages of FIA are speed (up to 300 samples per hour), simplicity and reduced cost of the apparatus. The very simplicity of this technique constitutes its main disadvantage. Very high sampling rates can be achieved for the simpler analytical procedures, but the technique cannot be applied to the more lengthy and elaborate operations, for which the air-segmented approach is better suited. There is still a lot of scope for innovation in both techniques.

14.4 Other CF techniques

Solvent extraction is a well-established technique in CF systems, using ordinary mixing coils followed by the appropriate separator fitting. Fittings, with special plastic inserts, are available according to whether the organic phase is light or heavy, and which of the two separated phases is required. The extraction solvent is not always suitable for direct injection on to an HPLC column. In these cases, an EDM (R) [evaporation to dryness module] is available in which the sample is continuously evaporated on to a moving 'wire' and then re-dissolved into an alternative solvent.

The high cost of specific enzymes has often precluded their use as continuously pumped reagents, but immobilized enzymes, coated on the inner surface of CF fittings, can be used for a variety of determinations including glucose (glucose oxidase) and penicillin (ß-lactamase). Both air-segmented and FIA systems have been used.

A stream splitter can be used to divide a sample stream equally between two manifolds. By this means, two separate determinations can be carried out using a single sample probe. If the two manifolds are identical, an essential reagent can be substituted in one manifold to make this stream an effective 'blank'. By leading the two streams to the respective cells in a twin-channel detector, simultaneous correction can be obtained by subtracting the two signals. It is necessary to accurately synchronize the two channels: this is easily accomplished by sampling a liquid which will give an equal response in both channels. By adjusting the lengths of transmission tubing feeding the two flow cells, a null signal will be obtained when the two channels are in phase. The mercury-catalysed 'imidazole' assay for ß-lactam antibiotics can be conducted in this way by simply omitting mercuric chloride from the 'blank' channel. A very wide range of formulated products can be assayed without interference from excipients.

A variety of switching-valve arrangements has been used to monitor the progress of tablet dissolution-rate tests by linking up to six dissolution vessels to a CF analysis system.

In this short chapter, it has only been possible to give an outline sketch of CF analysis, and whole books [16–19] have been written on the subject. It is hoped that a few appetites may have been whetted and that those previously unfamiliar with this technique be persuaded to dig deeper.

References

1. Skeggs, L.T. (1957) *Amer. J. Clin. Pathol.*, **28**, 311.
2. Aris, R. (1956) *Proc. Roy. Soc. A*, **235**, 67.
3. Sutter, A., Gardanier, S.A. and Spooner, G.H. (1970) *Amer. J. Clin. Pathol.*, **54**, 341.

4. Snyder, L.R. and Adler, H.J. (1976) *Anal. Chem.*, **48**, 1017.
5. Snyder, L.R. (1976) *Advances in Automated Analysis, 7th Technicon International Congress*, Vol. 1, p. 76. Technicon Instrument Corp., Tarrytown, New York.
6. Thiers, R.E., Cale, R.R. and Kirsch, W.J. (1967) *Clin. Chem.*, **13**, 145.
7. Pennock, C.A., Moore, G.R., Collier, F.M. and Barnes, I.C. (1973) *Med. Lab. Tech.*, **30**, 145.
8. Thiers, R.E., Reed, A.H. and Delander, K. (1971) *Clin. Chem.*, **17**, 42.
9. Walker, W.H.C. and Andrew, K.R. (1974) *Clin. Chim. Acta*, **57**, 181.
10. Snyder, L.R., Levine, J., Stoy, R. and Conetta, A. (1976) *Anal. Chem.*, **48**, 942A.
11. Neeley, W.E., Wardlaw, S.C. and Swinnen, M.E.T. (1974) *Clin. Chem.*, **20**, 78.
12. Neeley, W.E., Wardlaw, S.C., Yates, T., Hollingsworth, W.G. and Swinnen, M.E.T. (1976) *Clin. Chem.*, **22**, 227.
13. Walker, W.H.C., Townsend, J. and Kean, P.M. (1972) *Clin. Chem. Acta*, **36**, 119.
14. Caryle, J.E., McLelland, A.S. and Fleck, A. (1973) *Clin. Chem. Acta*, **46**, 235.
15. Betteridge, D. (1978) *Anal. Chem.*, **50**, 832A.
16. Furman, W.B. (1976) *Continuous Flow Analysis*. Marcel Dekker, New York.
17. Coakley, W.A. (1981) *Handbook on Automated Analysis*. Marcel Dekker, New York.
18. Foreman, J.K. and Stockwell, P.B. (1975) *Automated Chemical Analysis*. Ellis Horwood, Chichester.
19. Ruzicka, J. and Hanson, E.H. (1981) *Flow Injection Analysis*. Wiley, New York.

Glossary

Terms and abbreviations used in absorption spectrometry

Full definitions and examples of most of the terms will be found by reference to the index. In general, SI units are employed, but in several instances older units are still used by most spectroscopists. Some abbreviations used in the book are not listed here, but will be found in the Index.

A   *See* Absorbance.

$A_{1\,cm}^{1\%}$ *See* $E_{1\,cm}^{1\%}$

Å   *See* Ångstrom unit.

Absorbance   Quantity expressing the absorption of radiation by a solution at a specified wavelength. It is given by:

$$A = \log 1/T = -\log T$$

and is linearly related to the pathlength and concentration of the solution. It is dimensionless, but is expressed in absorbance units (A) so that a solution of $T = 0.1$ has an absorbance of 1 A.

Absorption   The process by which radiation is attentuated on passing through a substance. The term implies that the radiant energy is converted into some other form, e.g. heat, fluorescence, etc. as distinct from losses by scattering or refraction.

Absorption band   *See* Band.

Absorption spectrum   A plot of the absorption of radiation by a sample against the wavelength of the radiation.

Absorptivity   The absorbance of a solution of a compound in unit concentration measured in unit pathlength at a specified wavelength. *See* Molar absorptivity.

Ångstom unit (Å)   A unit of wavelength, now rarely used in absorption spectrometry. $1\,\text{Å} = 0.1\,\text{nm} = 10^{-10}\,\text{m}$.

b   *See* Pathlength.

Band   A general term describing a maximum in a plot of some quantity against wavelength. An absorption band is a broad maximum in an absorption spectrum and may comprise a number of minor peaks. An emission band might be the profile of an intensity versus wavelength plot of the spread of wavelengths leaving a monochromator.

Band pass *See* Passband

Beer's law (Beer–Lambert–Bouguer law) This relates the absorbance of a solution to the pathlength of the cell and the concentration of the solute. The absorbance of a solution at a specified wavelength is given by:

$$A = \epsilon bc$$

where ϵ is the molar absorptivity at that wavelength, b is the pathlength in cm and c is the molar concentration.

Blazed grating A diffraction grating made with a particularly high reflectivity in a particular spectral region.

c Velocity of light.

cm^{-1} *See* Wavenumber.

cps Cycles per second; *see* Hertz.

Cell A container to hold solutions for the measurement of their absorption spectra. This should have two parallel optical windows fixed at a specified distance apart.

Coated optics The coating of optical components to improve their optical performance and protect them against atmospheric attack.

Cuvette *See* Cell.

Dark current The background signal from a detector due primarily to the thermal emission of electrons.

Derivative spectra A plot of the first, second or higher derivative of an absorbance spectrum with respect to wavelength in order to correct for background absorption and to accentuate minor features in the spectrum.

Detector The device in a spectrometer which measures the intensity of light transmitted by a sample.

Deviation The displacement of the spectrometer measuring beam due to optical defects in the cell, or in its alignment.

Difference spectrum Small changes in the absorbance spectrum of a sample can be more readily detected if the spectrum is measured using the original sample as a reference solution.

Diffraction grating A device used to resolve radiation into its component wavelengths by an interference process. *See* Grooves.

Dispersion The power of a prism or grating to separate radiation of different wavelengths.

Double-beam spectrometer An instrument in which the measuring beam is split into two equivalent paths, one passing through the sample and one through a reference cell.

Dual-wavelength spectrometry Measurement of the absorbance of a sample at two wavelengths simultaneously in order to compensate for background absorption, etc.

$E_{1\,cm}^{1\%}$ An expression of the absorptivity of a solute of unknown molecular weight. It is the absorbance of a 1% w/v solution of a compound measured in 10 mm pathlength. It is related to molar absorptivity by:

$$E_{1\,cm}^{1\%} = 10 \times \epsilon/M$$

ESW *See* Effective spectral slitwidth.

Effective spectral slitwidth (ESW) The band of wavelengths emerging from a monochromator, expressed as the bandwidth at half-peak height.

Electromagnetic radiation Radiation which can be regarded as wave motions of characteristic wavelength, including γ-radiation, X-rays, UV, light, IR and radio waves.

Emission The process by which electromagnetic radiation is radiated from a substance.

Emission line Emission of radiation in a very narrow wavelength band characteristic of the output of gas discharge lamps.

Excited state A molecule that has absorbed UV or VIS radiation is said to have been raised from its ground state to an excited state.

Extinction *See* Absorbance.

Extinction coefficient *See* Molar absorptivity.

Far stray-light The component of the monochromator stray-light lying at wavelengths well away from the passband. This may be of low intensity, but can be a major factor in the problem of instrumental stray-light.

Far-ultraviolet Radiation at the shortwave end of the UV region. There are no agreed limits, but the far-UV is generally taken to lie between 190 and 250 nm.

First order *See* Grating order; Derivative spectra.

Flow cells A cell designed for the measurement of a flowing stream of solution.

Fluorescence A process in which radiation absorbed by a molecule is instantaneously re-emitted as UV or VIS radiation. This is usually of longer wavelength than the exciting radiation.

Fluorimeter An instrument specifically intended for the measurement of fluorescence.

Frequency (ν) A specification of electromagnetic radiation by the number of waves per unit time. It is related to wavelength by:

$$\nu = \lambda/c \text{ Hz}$$

where λ is the wavelength in m and c is the velocity of light in m s^{-1}.

Gaussian Most absorption bands when plotted on a frequency scale have a shape equivalent to a Gaussian or normal distribution, and can be fitted by:

$$A = A_m \exp - 2.773((\nu - \nu_m)/w)^2$$

where A is the absorbance at frequency ν, A_m is the absorbance at the maximum frequency ν_m, and w is the width of the band at $0.5 A_m$. See Lorentzian.

Grating *See* Diffraction grating.

Grating order A diffraction grating will reflect a given wavelength of light at more than one angle. Most of the incident energy undergoes first-order reflection, the second-order reflection is weaker, etc.

Grooves A diffraction grating works by the interference of light rays reflected from the faces of grooves ruled in its surface. The spacing of these grooves determines the optimum wavelength range for the grating.

Hz *See* Hertz.

Hertz Unit of frequency equivalent to one wave or cycle per second.

Holographic grating A diffraction grating produced photographically by the interference of beams of light.

IR *See* Infra-red.

ISL *See* Instrumental stray-light.

Incident beam The radiation directed into the entrance window of the cell in the spectrometer.

Infra-red The region of electromagnetic radiation of wavelength greater than the red end of the VIS region, i.e. from 800 nm to 1000 μm.

Instrumental stray-light (ISL) The overall effect of stray-light in a spectrometer manifested as detector output. Thus, it is the signal generated in the detector by all wavelengths outside the monochromator passband that reach it.

Isosbestic point If a sample contains two compounds in equilibrium, then at any wavelength at which their molar absorptivities are the same, the absorbance of the sample will be independent of the relative amounts of the two species. When the spectra of a series of mixtures of differing compositions are plotted, these will be seen to cross at these wavelengths, forming isosbestic points.

Light *See* Visible light.

Lorentzian Some absorption bands when plotted on a frequency scale have a shape equivalent to a Lorentzian or Cauchy distribution and can be fitted by:

$$A = A_m / \{1 + [2(\nu - \nu_m)/w]^2\}$$

where A is the absorbance at frequency ν, A_m is the absorbance at the maximum frequency ν_m, and w is the width of the band at $0.5 A_m$. *See* Gaussian.

Luminescence A general term for the emission of UV or VIS radiation from a molecule. In most cases, it can be regarded as the sum of the fluorescence and the phosphorescence emissions.

MSL *See* Monochromator stray-light.

mμ Millimicron; *see* Nanometre.

Manual spectrometer An instrument without a wavelength drive mechanism.

Mask An opaque screen with an aperture designed to limit the cross-section of a light beam.

Microcell A cell with a narrow chamber requiring the minimum volume of sample for a given pathlength.

Micron Micrometre (μm), 10^{-6} m.

Millimicron Nanometre (nm), 10^{-9} m.

Molar absorptivity (ϵ) The absorbance at a specified wavelength of a solution of a compound of unit molar concentration measured in a 10 mm pathlength. It has dimensions of $M^{-1} cm^{-1}$.

Monochromator stray-light (MSL) Radiation emerging from a monochromator of wavelengths outside the passband to which the monochromator is set.

NBW *See* Natural bandwidth.

NIR *See* Near-infra-red.

Natural bandwidth (NBW) The width of an absorption band of a particular substance, measured at half-peak height.

Near-infra-red The part of the infra-red region closest to the VIS region, generally taken to be between 800 nm and 2 µm.

Near stray-light The component of the monochromator stray-light lying within a few nanometres of the monochromator passband.

Near-ultraviolet The part of the UV region closest to the VIS region. There are no agreed limits, but the range is generally taken to be 250–400 nm.

OD *See* Optical density.

Optical density This is a general term for the absorbance of any material: absorbance should only be applied to solutions.

Passband The band of wavelengths emerging from a monochromator, generally centred on the indicated wavelength.

Pathlength The thickness of a sample solution traversed by the beam as it passes through a cell.

Phosphorescence A process in which radiation absorbed by a molecule is re-emitted as UV or VIS radiation, generally of longer wavelength than the exciting radiation. Unlike fluorescence, there is a delay between absorption and emission.

Photon A quantum of UV or VIS radiation.

Phototube A vacuum tube (valve) detector.

Photovoltaic detector A detector which generates a voltage in response to light.

Quantum A fundamental unit of radiation. The energy of a quantum is related to the frequency of the radiation by:

$$E = h\nu$$

where h is Planck's constant.

Radiation *See* Electromagnetic radiation.

Reference cell A cell identical to that used for the sample solution, but containing only solvent. The difference between the absorbances of sample and reference cells is thus a true measure of the absorbance of the solute.

Resolving power The ability of an instrument to distinguish between two closely-spaced maxima.

SSW *See* Spectral slitwidth.

Sampling cell   A cell designed to measure a number of discrete samples in succession, each sample displacing its predecessor.

Scan   To drive a spectrometer through its wavelength range in order to measure an absorption spectrum.

Scatter   *See* Scattered light.

Scattered light   Radiation which is reflected or refracted out of the measuring beam during its passage through the spectrometer. This may occur due to optical defects, dust or contamination of the optical surfaces in the instrument, or due to particles or inhomogeneities in the sample solution.

Second order   *See* Grating order; derivative spectra.

Semi-micro cell   A cell with chamber of reduced width and thus requiring less sample solution to fill it.

Single-beam spectrometer   An instrument with a single optical path. This means that for each wavelength setting, the sample cell must be moved out of the beam so that the transmission scale can be set to unity.

Sipper system   A manual or electric pump system used to draw sample solutions into a sampling cell.

Slew rate   A term used to describe both the rate of wavelength scan of a spectrometer and the rate of movement of the recorder pen.

Slit   An opaque screen with a fixed or adjustable aperture designed to limit the width of a light beam.

Slitwidth   The width of the aperture of a slit. Monochromator slits determine both the amount of light and the spread of wavelengths transmitted by the monochromator.

Spectral line   Generally synonymous with emission line, but can be used for very narrow absorption maxima.

Spectral slitwidth (SSW)   The range of wavelengths emerging from a monochromator when it is set to a particular wavelength and slit opening. Effective spectral slitwidth is the preferred means of expressing this.

Spectrometer   An instrument for measuring the transmittance or absorbance of a solution at different wavelengths.

Spectrophotometer   *See* Spectrometer.

Spectrum   *See* Absorption spectrum.

Stray-light   Radiation present in a spectrometer beam of wavelengths outside the monochromator passband. This may be due to optical defects, dust, etc.

T See Transmittance.

Transmission The process by which radiation passes through a material. It therefore represents radiation that is not absorbed, scattered or otherwise dispersed by the material.

Transmittance The proportion of light that is transmitted by a sample:

$$T = I/I_0 = 10^{-A} = \mathrm{antilog}(-A)$$

where I_0 and I are the intensities of the light falling on the sample and that emerging from it. Thus, T lies in the range from 0, for an opaque material, to 1 for a transparent one. T has no units, but is sometimes expressed as a percentage, i.e.

$$T\% = 100 \times I/I_0$$

UV See Ultraviolet.

UV–VIS spectrometer An instrument designed to operate through the UV and VIS regions, i.e. from 180 to 800 nm.

Ultraviolet (UV) The region of the electromagnetic spectrum lying at shorter wavelengths than the blue end of the visible region, and generally taken to be between 100 and 400 nm.

VIS See Visible light.

Vacuum ultraviolet The shortwave part of the UV region where oxygen and other gases absorb strongly, lying between 100 and 180 nm. Spectrometers for this region must be housed in evacuated chambers.

Visible light The region of electromagnetic radiation that can be seen by the human eye, and generally taken to extend from 400 nm (violet) to 800 nm (red).

Wavelength (λ) A measure of electromagnetic radiation, being the length of the waves associated with the radiation. In the UV–VIS region, these waves are very small being of the order of 10^{-7}–10^{-6} m in length; about 150 waves of green light would span the thickness of this page.

Wavenumber ($\bar{\nu}$ cm^{-1}) An alternative measure of frequency, expressed as the number of waves cm^{-1}. This can be derived from the wavelength without knowledge of the velocity of light:

$$\bar{\nu} = 10^7/\lambda \, \mathrm{cm}^{-1}$$

where λ is the wavelength in nanometres.

Window The optical faces of a cell.

Working area That part of the window of a cell that is up to optical specification and through which the measuring beam can pass without risk of interference with the walls or floor of the cell, or with the meniscus of the sample.

Zero order In discussing derivative spectra, zero order is used to describe the fundamental absorption spectrum.

α Sometimes used for absorptivity or molar absorptivity.

ϵ Molar absorptivity.

ϵ_{max} The maximum absorptivity of an absorbance band.

λ Wavelength.

λ_{max} The wavelength of maximum absorbance of an absorption band.

μm Micrometre.

ν Frequency.

$\bar{\nu}$ Wavenumber.

Appendix: Selected publications in the Bulletins of the UV Spectrometry Group 1949–1983

Introduction

The originally titled Photoelectric Spectrometry Group Bulletin (PSGB) ran to 20 issues from April 1949 to July 1972. This was continued under the name of the UV Spectrometry Group Bulletin (UVSGB) for 15 issues from August 1973 to December 1983. These 35 slim volumes span the formative years of modern spectrophotometry, and it is appropriate to record these contributions as part of our 50th anniversary. The following extract is from the very first PSGB editorial by the Group's first chairman Dr J.R. Edisbury.

Genesis

'To avoid possible future speculation and the growth of improbable legends concerning the origin of the Photoelectric Spectrometry Group, let the authentic details now be put very briefly on permanent record. In July, 1947, I received from the Thornton Research Centre of the Shell Refining and Marketing Co., Ltd., a slim volume entitled "The Beckman Photoelectric Spectrophotometer (Model DU) with ultra-violet accessories". Notes on design, operation and maintenance, compiled by three present members of our Group, Beaven, Bentley and Sutciffe. Although slim, the volume was obviously packed with erudition and likely to be of extreme interest and practical value to Beckman users. In my letter of thanks, I casually and incautiously mentioned that 'a Beckman Discussion Panel ... for the informal interchange of information ... might be worth-while. Do you think there are sufficient local people interested?" An informal meeting of workers in the Liverpool area supported the idea, and a short list of "local interest" was compiled, several names from more distant parts were added, the scope was widened to cover other instruments of similar type, our Hon. Secretary was unobtrusively bludgeoned into taking office, and in June,

1948, a circular and questionnaire were sent out to 84 prospective members. This was followed by a Notice of the Inaugural Meeting to about 120, the increase being the result of fresh suggestions from those already canvassed. Paid up membership now (April, 1949) stands at about 130. Sixty-four attended the Inaugural Meeting at Cambridge on July 16th. Members came from as far afield as Bucksburn, in Aberdeenshire, and Poole, in Dorset. In his chair at the bottom of the Chemistry Lecture Room No. 1 in Pembroke Street, seemingly almost underneath the towering tiers of members, the Chairman felt rather like Exhibit "A" at the wrong end of a microscope. The agenda was perhaps over-ambitious and two controversial items had to be shelved for future discussion: the Chairman had imagined in his simplicity that a Photoelectric Spectometry Group would wish to concern itself primarily with photoelectric spectrometers and accessories therefore; also he was frankly unprepared for quite so many people to express quite so many different opinions on affiliation. However, the main items were amicably settled, a well-representative Committee elected and the tea-break reached only five minutes behind schedule.'

References

A.1 Absorbance and wavelength standards

Knowles, A. (1978) Standards Working Party report meeting. *UVSGB*, **6**, 56.

Burgess, C. (1977) Monitoring the performance of UV-visible spectrophotometers. *UVSGB*, **5**, 77.

Douglass, S.A. and Emary, R.J. (1977) Standardization of temperature, absorbance and wavelength measurements in UV–VIS spectrophotometry. *UVSGB*, **5**, 85.

Popplewell, B.P. (1977) The calibration of neutral density filters. *UVSGB*, **5**, 90.

Knowles, A. (1977) A new commercial system of solutions for spectrophotometer checks. *UVSGB*, **5**, 94.

Clarke, F.J.J., Davis, M.J. and McGivern, W. (1977) Transmittance standards from NPL. *UVSGB*, **5**, 104.

Sharpe, M.R. (1975) Some investigations into the use of filters for calibrating UV-VIS spectrophotometers. *UVSGB*, **3**, 57.

Johnson, E.A. (1967) Potassium dichromate as an absorbance standard. *PSGB*, **17**, 505.

Everett, A.J., Young, P.A. *et al.* (1965) Official report on the PSG collaborative test of recording spectrophotometers. *PSGB*, **16**, 443.

Tarrant, A.W.S. (1965) Some comments on the findings of the PSG collaborative test. *PSGB*, **16**, 458.

Glenn, A.L. (1965) Further comments on collaborative tests. *PSGB*, **16**, 464.

Neal, W.T.L. (1956) Differential absorptiometry. *PSGB*, **9**, 204.

Ketelaar, J.A.A., Fahrenfort, J., Haas, C. and Brinkman, G.A. (1955) The accuracy and precision of photoelectric spectrophotometers. *PSGB*, **8**, 176.

Morton, R.A. (1951) Collaborative test on potassium dichromate – introductory remarks. *PSGB*, **4**, 65.

Gridgeman, N.T. (1951) Statistical analysis: the accuracy and precision of photoelectric spectrophotometry. *PSGB*, **4**, 67.

Harding, H.G.W. (1951) Instrumental aspects: precautions necessary for accurate measurements of O.D. standards. *PSGB*, **4**, 79.

Lothian, G.F. (1951) Notes on future spectrophotometric tests. *PSGB*, **4**, 86.

Edisbury, J.R. (1950) Further comments on the P.S.G. collaborative test with potassium nitrate. *PSGB*, **2**, 32.

Edisbury, J.R. (1949) Collaborative test: relative readings on twenty-eight Beckman spectrophotometers. *PSGB*, **1**, 10.

Irish, D. (1983) Use of rare earth oxide solutions for wavelength calibration in UV–VIS spectrophotometry. *UVSGB*, **11**, 39.

Verrill, J.F. (1983) A re-evaluation of metal film on silica neutral density filters. *UVSGB*, **11**, 30.

Carr, G.P. (1981) A pharmacopoeial view of absorption standards. *UVSGB*, **9**(1), 64.

Clewes, B.N. (1979) Wavelength calibration of UV-Visible spectrophotometers using Legendre polynomial functions. *UVSGB*, **7**, 35.

A.2 Stray-light

Knowles, A. (1978) Stray-light in grating monochromators. *UVSGB*, **6**, 84

Sharpe, M.R. and Irish, D. (1976) Stray-light in grating spectrophotometers. *UVSGB*, **4**, 51.

Hartfee, E.F. (1963) Stray-light in ultraviolet spectrophotometers: the need for a standard criterion. *PSGB*, **15**, 398.

Lothian, G.F. (1956) Effects of finite slitwidth and stray radiation in differential absorptiometry. *PSGB*, **9**, 207.

Cannon, C.G. (1955) Anomalous slitwidth effect in differential absorptiometry. *PSGB*, **8**, 201.

Anon. (1952) Proposed collaborative stray-light test. *PSGB*, **5**, 119.

Collins, F.D. (1951) Notes and observations: a postscript on stray-light. *PSGB*, **4**, 96.

Perry, J.W. (1950) Sources and treatment of stray-light in spectrophotometry. *PSGB*, **3**, 40.

Donaldson, R. (1950) Measurement of stray-light by double monochromator. *PSGB*, **3**, 45.

Martin, A.E. (1950) Stray-light in infrared spectrometry. *PSGB*, **3**, 50.

A.3 Cells

Knowles, A. (1978) Cell Working Party report meeting. *UVSGB*, **6**, 54.
Goddard, D.A. (1976) The cleaning of vitreous silica cells. *UVSGB*, **4**, 19.
Archer, M.S. (1954) Techniques for using absorption cells in ultraviolet spectro-photometry. *PSGB*, **7**, 160.

Abbreviations: *UVSGB: UV Spectrometry Group Bulletin; PSGB: Photoelectric Spectrometry Group Bulletin.*

Index

absorbance
  accuracy checks, 141
  definition of, 232
  measuring, 165–76
absorbance standards
  crossed polarizers, 73–4
  glass filters, 69
  light addition methods, 77
  metal screens, 74–5
  sector disc, 75
  solid, 69–80, 141
  solutions, 54–68
absorbance standards
  thin film filters, 75–7
absorption band, 149
absorption spectrum
  aniline, 153
  anthracene, 154
  benzene, 135
  carbonyl compounds, 155
  cobalt ammonium sulphate, 62
  cytosine, 157
  didymium glass, 113
  ethanol, 88
  ethinyl oestradiol, 211
  holmium glass, 112
  holmium oxide solution, 137
  ethylbenzene, 198
  isophalphic acid, 216
  maleic acid, 87
  McCrone filter, 138–9
  naphthacene, 154
  naphthalene, 154
  nicotinic acid, 61
  potassium chromate, 55
  potassium dichromate in acid, 54
  potassium dichromate in alkali, 58
  potassium hydrogen phthalate, 60
  potassium nitrate, 61
  riboflavin, 187
  teraphthalic acid, 216
  toluene, 132, 151
  xylene, 198
accreditation, 125–7
accuracy measurements, 180
American Society for Testing and Materials, 136
ammonium cobalt sulphate, *see* cobalt ammonium sulphate
aniline, 150
anthracene, 154
arc lamps, *see* discharge lamps
auto analyser, 221
automated sample handling, 220–28
Ayers plot, 184

background correction, 190
bandwidth, definition of, 161
bathochromic shift, 155
beam instrumental considerations, 44–7
Beer–Lambert law, 5, 158
benzene, 132, 150–52
benzene vapour absorption spectrum, 135, 152

cells, 18–42, 44–50, 167–9

angular deviation, 47
cleaning, 37, 179
cylindrical, 46
design of, 19–23, 34–41
flow through, 28, 40
for fluorescence, 31, 38–9
grades of, 18, 27
handling, 8, 37, 167–9
holders, 47
labelling of, 41–2
matching, 48
materials for, 28–31
methods of construction, 34–6
micro, 25, 46
normal, *see* standard
optical specification, 32–4
rectangular, 23–5, 46
reflection at windows, 7, 160
sampling, 23
semi-micro, 25
standard, 24–6
storage of, 180
transmission of windows, 31–2
working area, 22, 24–5
centrifugal analysis, 220
chance neutral filters, 70
checks of instrument performance, 130–42
CITAC, 121
cobalt ammonium sulphate, 61–3
continuous flow systems, 220–28
   air segmented, 221
   pumps for, 224
cross-filter technique for wavelength calibration, 118
curve fitting, 202
cuvettes, *see* cells

data smoothing, 193
deconvolution of spectra, 201–4
densitometry, 217
derivative spectra, 206–11
deuterium emission lines, 111–12
dichromate ion, 53
didymium glass, 110

difference spectroscopy, 211
diode array instruments, *see* spectrometers diode array
discharge lamps for wavelength calibration, 108–9, 132–3
dual wavelength spectroscopy, 214
dyes as absorbance standards, 63

effective spectral slitwidth, 162
electromagnetic radiation, 147
electronic transitions, 152
excited state, 149
extinction coefficient, *see* molar absorptivity

FDA, 122
filters
   as absorbance standards, 69–73
   as wavelength standards, 109
   to reduce stray light, 88
first-order derivative, 207
flow injection analysis (FIA), 227
Fourier transform, 43, 204
fused quartz, 30

GLP, 122
GMP, 121

high performance liquid chromatography, 51
holmium (III) perchlorate, 112, 134
holmium oxide glass, 110–12
HPLC, *see* high performance liquid chromatography
hypsochromic shift, 157

IEC Guide 25 (ISO 17925), 123, 125
ILAC, 124
instrumental stray light (ILS), *see* stray light
interference filters, 117
ISO 17925, *see* IEC Guide 25
ISO 9000, 123, 125

Index

Laboratory of the Government Chemist, 66
Langdales Sap Green Food Colouring, 63
least-squares smoothing, 195
Legendre polynomials, 193
Lorentzian absorption bands, 208

masks, *see* spectrometers, beam making
matrix compensation methods, 212
matrix rank analysis, 199
McCrone wavelength standard, 113, 134
mercury discharge lamps, 109, 132
molar absorptivity, 158
monochromator stray light (MSL), 82
monochromators
   resolution of, 130–31
   wavelength calibration, 108–18
Morton and Stubb's correction, 190–92
moving average, 195
multicomponent analysis, 196

National Bureau of Standards (NBS), *see* National Institute of Standards and Technology
National Institute of Standards and Technology (NIST)
   calibrated cells, 113
   calibrated glass filters, 70–73
   calibrated thin film filters, 75–7
   composite solutions, 65
   standard reference materials, 65–6
   web site, 66
National Physical Laboratory (NPL)
   calibrated glass filters, 70–73
   calibrated thin film filters, 75–7
   McCrone wavelength standard, 113, 134

neodymium oxide glass, *see* didymium glass
neon emission lines, 110
neutral density filters as absorbance standards, 70
nicotinic acid, 60

optical density, 237

Photoelectric Spectrometry Group, 241
Plank's constant, 149
plastic in cell construction, 30
potassium chromate, 58
potassium dichromate
   in acid solution, 53–7, 141
   in alkaline solution, 57–8
potassium hydrogen phthalate, 59
potassium iodide, 91
potassium nitrate, 61
praesodymium oxide, *see* didymium glass
precision of measurement, 180–86

qualification of instruments, 127–9
quartz, *see* fused quartz; synthetic fused silica

radiation
   absorption of, 4
   attenuation of, 5
   frequency of, 4
   intensity of, 4
   near-UV region, 147
   visible region, 147
   wavelength of, 4
Rayleigh scattering, 208, 214
reflection losses at cell windows, 6–7
resolution, *see* monochromator resolution
Ringbom–Ayres plot, 184

samarium (III) perchlorate, 116
sample handling, 8, 166, 186–9

scan rate, choice of, 171
scattered light, 160
Schott neutral filters, 70–73
second-order derivative, 206–11
silica, *see* synthetic fused silica; fused quartz
slitwidth
    definition of, 162
    optimum, 11, 131, 172
sodium iodide, 82
solid samples, 186
solutions, preparation of, 166
solvent
    choice of, 14, 165
    effect on spectrum, 155
    transmission of, 14
solvent cut-off wavelength
spectral bandwidth (SBW), *see* slit width
spectral stripping, 201
spectrometer cells, *see* cells
spectrometers
    absorbance accuracy checks, 53–67, 140
    beam dimensions, 44
    beam divergence, 44–7
    beam masking, 46
    calibration by filters, 69–73, 75
    calibration by light addition methods, 77
    cell positioning, 47–50
    checking procedures, 130–42
    design of, 43
    diode array, 50
    linearity checks, 140
    optimum absorbance range, 10
    optimum slitwidth, 172
    self-calibrating, 50
    stray light in, *see* stray light
    wavelength calibration, 108–115

wavelength resolution, 130
spectrophotometers, *see* spectrometers
spectroscopic terms, 230–38
standard reference materials (SRM), 65, 140
stray light, 12, 16, 81–106
    definition of, 81
    errors due to, 13, 84
    measurement of, 89–105, 135–40
    near, 163
    origins, 85
    reducing the effects of, 85
stray radiant energy (SRE), *see* stray light
synthetic fused silica, 30

thin layer chromatography, 218
transfer pipette, 169
transmission of cell windows, 31
transmittance, definition of, 5
turbid samples, 187

UKAS, 125
ultraviolet radiation, 147
Ultraviolet Spectrometry Group (UVSG), formation of, 239

vacuum ultraviolet, 148
visible light, 148

wavelength, 4
wavelength calibration
    by absorption standards, 109–15
    by emission standards, 108
    by interference methods, 117
    for wide-band instruments, 115
wavenumber, 4, 147
working area, *see* cells, working area